Early
Childhood and

Also available from Bloomsbury

Early Childhood Studies, Ewan Ingleby

Early Childhood Theories and Contemporary Issues,
Mine Conkbayir and Christine Pascal

Reflective Teaching in Early Education, edited by Jennifer Colwell

Early Childhood and Neuroscience

Theory, Research and Implications for Practice

Mine Conkbayir

Bloomsbury Academic
An imprint of Bloomsbury Publishing Plc

BLOOMSBURY
LONDON · OXFORD · NEW YORK · NEW DELHI · SYDNEY

Bloomsbury Academic

An imprint of Bloomsbury Publishing Plc

50 Bedford Square	1385 Broadway
London	New York
WC1B 3DP	NY 10018
UK	USA

www.bloomsbury.com

BLOOMSBURY and the Diana logo are trademarks of Bloomsbury Publishing Plc

First published 2017

© Mine Conkbayir, 2017

Mine Conkbayir has asserted her right under the Copyright, Designs and Patents Act, 1988, to be identified as Author of this work.

British Library Cataloguing-in-Publication Data

A catalogue record for this book is available from the British Library.

ISBN: HB: 978-1-4742-3191-6
PB: 978-1-4742-3190-9
ePDF: 978-1-4742-3192-3
ePub: 978-1-4742-3193-0

Library of Congress Cataloging-in-Publication Data

A catalog record for this book is available from the Library of Congress.

Typeset by Fakenham Prepress Solutions, Fakenham, Norfolk NR21 8NN
Printed and bound in Great Britain

For Delilah, canım benim

Contents

List of Figures

Foreword

Relating the Miracle of Young Life to Mysteries of the Growing Brain

Early years author and trainer Mine Conkbayir is currently co-writing two books with speech and language therapist Michael Jones. She was previously head trainer of practitioners for the London Early Years Foundation, and is the author, with Chris Pascal, of a book about *Early Childhood Theories and Contemporary Issues*, published in 2014. They reviewed old and new theories of the natural abilities of young children, and how these talents respond to thoughtful teaching that welcomes their creativity.

The child's playful zest for a life of learning, and how they wish to share it, is again the principal topic of Mine's second book. Here she also makes an assessment of what science of the brain and of its growth in different environments may contribute to her understanding of early development as a teacher, or how certain scientific claims may mislead her. She is concerned that fragmentary and conflicting messages about neurones and their functions in the myriad changing tissues of the brain needs to be interpreted in an open-minded and critical way to assist well-being of a whole child alive in relations with parents and teachers.

Mine has long been interested in the development of services to give aid to children in stressful or neglectful relationships, families in poverty, and mothers or teachers who are depressed and unable to cope with and relate to childish vitality. She hopes new ideas from neuroscience might support her work, for both care and education.

As teachers and parents we need to appreciate how, in every human community, impulses for play inspire rituals of artful creativity and their celebration, and to consider how the source of this imaginative vitality born in our children, and treasured by the musicologist Jon-Roar Bjørkvold and the poet Kornei Chukovsky, may best be supported. Does our invention of machines to advance scientific attention to details of our biology, and the practice of technical projects that transform the environment, or that regulate industrial exploitation of resources, help or confuse awareness of the natural motives that generate all these enterprises? Does the ambitious world of adults searching for profits in knowledge and skills become toxic for the spirit of many children as the writer and counsellor Johann Christoph Arnold of the Bruderhof Community fears?

I am delighted to write an introduction to Mine's search for answers because I was trained as a young brain scientist to seek knowledge of the 'imagination', rather than the mechanism of intelligence and its learning. I studied with Roger Sperry, a pioneer of the theory that consciousness lives in the brain to move the body well. He led delicate studies of the growth of nerve circuits for action-with-awareness. He received a Nobel Prize for research on the functions of the two cerebral hemispheres which revealed them to be agents with different gifts that work together to guide movement of a coordinated conscious and emotionally coherent Self. There are complementary preferences and talents on the two sides of the mind in the brain, different ways of perceiving and using the environment, and for communicating and appreciating ideas with other people, and there are differences of 'talent', 'character' and 'personality' between us, but there are common motives and feelings that give aesthetic and moral values to all we do.

Sperry's work confirmed that imaginative projects for moving, and their emotional regulations, are generated in deep sub-cortical parts of the central nervous system. Following the birth of my first son, I have applied this idea in fifty years of research on how a baby is born ready to use their clever and imitative human nature to develop as a 'person in relation', seeking knowledge and skills that have meaning for other people. That is how we learn.

Neuroscience has a short history. Charles Darwin, who studied the communication of emotions in animal life and used the new invention of photography in the nineteenth century to show how rich emotional expressions are in young children, knew nothing of neural circuits. Charles Scott Sherrington, who established the foundations of modern brain research a few decades later, by studying how reflexive movements cooperate in creative ways, invented the word 'synapse' and investigated messages nerve cells transmit to each other. Following his 80th year he summarized his life's work in lectures, published as *Man on His Nature*. He explained how living organisms, even bacteria, grow and learn by exercising 'environment expectant' imagination, not just by memory of what has stimulated them. Sherrington discovered and named 'proprio-ception', the feeling of forces of the moving Self inside the body, and he described how the 'distance senses' of sight and hearing are used, for 'projicience' or 'reaching out' to anticipate the properties of the environment and of objects the body takes up. Everything we perceive has to be evaluated by 'affective appraisal', which refers to inner feelings of well-being estimated by 'viscero-ception'. All these functions of the brain that Sherrington distinguished make up one's self-consciousness, the feelings of which can be shared when emotions are communicated.

In his closing lecture, Sherrington advocated respect for affections of the mind that help 'altruism' in relationships, for cooperative life activities with sympathy with other persons' feelings, including therapeutic care of patients and education of the young. This is a philosophical neuroscience that goes beyond examination of details to take in the whole picture of organic, and human, mental life – the science needed for understanding of the vitality and affections of young children, and why they are so rewarding for loving parents, teachers and other companions who share their innocent life activities.

Mine takes pains to appreciate a comprehensive natural science of how the brain can animate a human mind and share consciousness. She distinguishes this from quests for scientific proof about cognitive appraisals of stimuli, or the damaging effects of a severely impoverished environment with neglect and abuse, both of which address questions that are too narrow in focus. Misguided and misleading views of how brain functions of intelligence originate, and the idea that they grow by 'plastic' response to the environment without evaluation by emotions in relationships, follow from what can be called 'reductive' or too pragmatic and rational prejudices about the initial state of the neural systems, and how they are adapted to be changed by learning. The error of this approach is explored by a young phenomenological philosopher, Barbara Goodrich, who is also interested in the veterinary care of animals. She supports the animated brain science of 'movement and time', rather than that which only considers abstract processes of the 'knowing and reasoning' mind. Her approach is close to that of the Chilean biologist and system theorist Humberto Maturana.

A case in point concerns the denial of imitation of intentions by babies. Belief that the child's mind grows by learning how to respond to stimuli, how to categorize objects, how to respond intelligently to the actions of other persons, and especially how to think in words, leads to the conviction that the inexperienced and speechless newborn cannot possibly perceive intentions in another's movements, and therefore cannot imitate a movement of expression, by eyes, head, face, mouth or hands, all of the organs uniquely adapted in humans for communication of interests and feelings. In actual fact, a newborn can, if it 'wants to', imitate all these movements.

Even more remarkable evidence has been obtained by thoughtful and 'respectful' examination of the responses of babies in the first hours after birth by a young medical psychologist, Emese Nagy. She showed that a newborn imitates to have a dialogue, to share intentions. After imitating, the baby watches provocatively to see if the other person will imitate them. Imitation is not just for learning – it is a sign that tries to confirm understanding of life between persons, that they are 'being in contact together'.

Since 1970, thanks to the work of pioneers like psychiatrist Daniel Stern and another medical doctor who became interested in developmental psycholinguistics, Margaret Bullowa, both of whom have actually studied mother–infant communication, appreciation has grown of the young infant as a person who can 'converse' with a parent or other affectionate partner a rhythmic exchange of movements that express interest and feelings that become memorable narratives and games as in baby songs. The baby shows talent for this proto-conversational play with self-conscious emotions of pride and shame, and in a few months wants to share meaningful tasks and tools, before talking about them. My work in this field, which began with Jerome Bruner at the Centre for Cognitive Studies in Harvard where he set out to change the theory of human cognition by focusing on how infants learn, has been encouraged and enriched by the findings of psychologist Vasudevi Reddy and musician Stephen Malloch, both of whom focus on intimate and adventurous sharing of feelings by subtle expression of self-consciousness in body movement.

The accidental discovery by the Italian neuroscientist Giacomo Rizzolatti and his colleagues of mirror neurones that enable imitation of actions may not fully explain human 'other awareness' with its rich emotional colours, but it certainly revolutionized ideas about the functions of the cerebral cortex and how its great capacity for storing experiences is motivated for convivial awareness. Mine gives attention to another revolutionary neuroscientific fact made clear by Stephen Porges, that the brainstem control of the movements and hormonal communications that govern inner visceral states and 'gut feelings' also shape the movements that communicate the emotions to other people. Similar concepts by Jaak Panksepp, who studies the 'affective consciousness' of the brainstem in mammals, and how it influences what the brain perceives and remembers, have given us an appreciation of how feelings of mindfulness invade all parts of the central nervous system and its dealings with the body. With this more imaginative and sympathetic brain science we understand better how a young child, even one who has the misfortune to be 'anencephalic', with severely undeveloped cerebral cortex, may be such a feelingful and sympathetic companion, ready for intense engagement of states of curiosity and well-being. It gives us confidence that there are hopeful forces of the mind that can help overcome very difficult problems in growing up.

Another topic of interest for teachers, about which there is vigorous debate, now concerns growth of the differences in feelings and awareness of the left and right hemispheres that Sperry's work illuminated over fifty years ago. It is found that in the period before school there is a major shift in growth between the two sides. In the first two or three years the right hemisphere, which Alan Schore shows is more sensitive and expressive for intimate affections of attachment, is growing more. After that the left hemisphere shows an accelerated growth that coincides with the acquisition of a vocabulary of speech. A psychiatrist, Iain McGilchrist, after making a thoughtful review of the knowledge we have of hemispheric differences in adults, proposed that modern industrial and technologically complex cultures, which depend on elaborate languages and artificial media, exhibit a weakening of aesthetic and intuitive appraisals of the environment, and especially of other persons' feelings and moral intuitions. They value too much what the left hemisphere can do. This idea is hotly contested by those who favour industrial development and its rationality, but it raises important implications for education at all levels. Cognitive classroom training of verbal and mathematical intelligence in preparation for employment, especially if it is imposed in the early years, may not aid development of imaginative enrichment of motives for cultural and interpersonal understanding. It may weaken the establishment of strong 'mutual expectancy' and affectionate respect between old and young, which Jerome Bruner has considered to be the foundation for cooperative learning and responsive teaching in any form of schooling.

The human brain has a uniquely complicated body to control, and a very rich set of emotions that regulate attachments and companionship, leading to sharing of stories that are passed on from generation to generation as the wisdom of culture in large and ancient societies. The Scottish developmental psychologist Margaret Donaldson distinguishes the 'human sense' we are all born with from the cultural 'common sense' which has to be learned as a community of knowledge, skills and creative ritual practices, including speaking a particular language. Psychologists who have attended to how these two ways of making sense of life appear and develop in sequence in early childhood have discovered radically new ways of thinking about how the human brain works. It learns by making close relationships that share both aesthetic and moral feelings about the pleasures and beauty of actions, and about the responsibilities of cooperation with the hopes and interests of other persons. The creation of a new developmental and educational philosophy has benefitted from studies of the spontaneous and creative activities of the whole brain that generate and regulate movements, and that store specific experiences in their context for future recognition, giving them qualities and affinities by 'affective appraisal'.

Mine Conkbayir's thought-provoking review of where we are now in our scientific understanding of early development of the child's mind and its biology is intended to share confidence with those who are happy in their work and glad to learn from it, and also to help mothers or teachers who are distressed or afraid and unable to be in touch with the innate enthusiasm of a young person. I am sure that gently showing up the cleverness and sensitivity of a baby can make a difference, either to a devoted teacher who feels the need for recognition by the educational establishment, or a parent experiencing the frustration of postnatal depression.

That is the idea that led the paediatrician T. Berry Brazelton to develop an entirely new practice of playful engagement with a newborn to encourage the first communications of mother and father with their baby, making them happy and proud. He extended the idea of humanizing medical care to accept that there are other special times in the natural growth of life when intimate and affectionate responses from companions support advances in collaboration that carry some risk of disappointment, and developed the 'touchpoints' approach, creating the Brazelton Touchpoints Centre which works 'to establish scalable and sustainable, low-cost, low-tech interventions that propel children's healthy development and build the internal capacity of – and strengthen the collaborative relationships among – families, parents, caregivers, providers, and communities' (http://www.brazeltontouchpoints. org/). This is a brilliant example of a multidisciplinary way to inform practices with up-to-date knowledge from psychology and brain science, but mainly with wide experience of success in improving and enriching activities of the young child as a self-confident and affectionate collaborator, first in the intimate relations of the family, and then in a community and its creative practices, traditions and beliefs.

Colwyn Trevarthen, Emeritus Professor,
Child Psychology and Psychobiology, University of Edinburgh, UK

Acknowledgements

I would like to thank Professor Eleanor Dommett and Colwyn Trevarthen, Emeritus Professor of Child Psychology and Psychobiology at the University of Edinburgh, who have given their considerable expertise and valuable time in helping with this book.

I am also grateful for the support of June O'Sullivan MBE, Chief Executive of the London Early Years Foundation (LEYF), Professor Chris Pascal and Play Therapist, Christian Bellissimo.

I hope that this book is a useful resource for early childhood practitioners in applying some of the exciting developments taking place in neuroscience.

Introduction

What to Expect and What Not to Expect from this Book

I have been inspired to write this book after many and varied interactions with early childhood professionals and neuroscientists concerning the potential of neuroscience to help inform early childhood practice. As an early childhood lecturer, researcher and author I am passionate about sharing knowledge and experience regarding how we can best utilize *relevant evidence-based* research, including that from the field of neuroscience, to help inform contemporary early childhood education and care.

When I have asked neuroscientists the question 'What can neuroscience tell us about *how* early experiences can impact brain development?' I have been disheartened to be repeatedly told 'very little', not least because this is not considered to be true by all those involved (Howard-Jones 2013; Rose and Abi-Rached 2013; Gopnik 2009; David et al. 2003). What I feel is more probable, based on these interactions – and the primary and secondary research that has been undertaken to write this book – is that there exists a reluctance to apply neuroscience fully to the area of childhood development. This perhaps exists to protect the integrity of the science; there is considerable fear of scientific research being 'overgeneralized' into areas which differ hugely from the context of the original research, and the findings subsequently getting lost in translation. However, even though this attitude may be held by many, it seems not to have prevented the overabundance of neuromyths (Satel and Lilienfeld 2013) being employed in the marketing of some consumer products, political campaigns and education programmes that make grand

and sometimes false claims apparently based on findings from neuro-science (which will be examined in Chapter 1). So if neuroscience is already applied to education, it is important to make sure it is applied appropriately, and perhaps even more important, that those at the frontline of education are well equipped to assess whether applications are valid before they adopt different practices.

Why this book is needed

There are few books at present which are dedicated to exploring the relationship between neuroscience and early childhood development and the practical implications of this relationship for early childhood education and care. Serving as an effective quick-reference tool mainly for early childhood students, lecturers and practitioners, this book is a user-friendly introduction to neuroscience and its potential to inform education and care in a range of early childhood settings. *This book is not intended to provide in-depth explanations of the brain imaging studies identified, but readers are encouraged to follow these up by referring to the bibliography.*

The concepts examined in this book are brought alive through the contri-butions of early childhood professionals, head teachers of early childhood settings, play therapists and parents. This book can be used by those who wish to refer to neuroscience when seeking alternative explanations concerning issues around child development and learning. Students, lecturers and practitioners alike can reflect on the broader implications of some of the topics explored. For example, discussions and debates can be had concerning epigenetics and the factors that play a role in the 'switching on and off' of genes, as well as the significance of the first 1001 days of life and the key influences which can start shaping development from conception. Questions concerning ethics and replicability in brain research as well as how inter-disciplinary working could be achieved may also arise for you when reading this book.

Early childhood students and practitioners are continually being encouraged and supported by lecturers and training professionals to use theoretical knowledge from the likes of Piaget, Bowlby and Vygotsky to help inform their planning of early childhood curricula, learning environ-ments and their interactions with very young children. This is not the case when it comes to using neuroscience, which can also be used to inform understanding of brain development in relation to their practice. It is time for neuroscience to finally be accepted as another way of theorizing about and understanding key issues concerning early childhood development and provision of education and care. This does not imply that understanding neuroscience is the only way to do this, but that it adds another, more contemporary dimension to our understanding.

As you make your way through this book, it will become apparent that the young brain is distinctively influenced and shaped by everyone and every-thing around it: perhaps most significant is the role of adults. Parents, primary carers, early childhood practitioners and teachers are fundamental in shaping the infant's emotional and social worlds and how they perceive themselves in these worlds. The overarching objective of this book is to help early childhood students and practitioners understand how to better provide for young children.

Who this book is aimed at

I hope that the information and ideas presented in this book can be taken and made meaningful in your practice. This might include in the nursery, pre-school or the primary school context, or even in your professional experience as an outreach worker or foster carer. Lecturers and students studying on level 3 early years courses or on a foundation degree in early childhood studies will also benefit from the concepts and discussions put forward in this book. Those with an interest in infant development with regard to the brain and how this is affected by environmental, socio-biological and neurobiological factors may find the issues examined here informative and, most importantly, practically useful. This book is intended to be a practical resource to build understanding and improve practice. Hopefully you will find it a challenging and exciting addition to your collection.

How this book is structured

The book contains six chapters, each following a similar format, which comprises one main discussion (as per the chapter title), with subheadings that give structure to each chapter. Questions are included for the reader to reflect on and answer, alongside case studies provided by various inter-disciplinary professionals, experts and parents, as well as suggestions for practice to help the reader to put the theories into a practical, real-life context. Where possible, these activities are best undertaken in groups, in order to generate debate and new ideas.

Throughout this book, you will notice that some words are written in bold and accompanied by this icon in the margin. [📖] These words can be found in the Glossary of Terms at the back of the book.

The term 'early childhood' is used interchangeably with the terms 'early years'. This is in recognition of the fact that readers in the United States use the term 'early childhood' more frequently than 'early years'. Early childhood in this book is taken to mean the period from conception up to five years, with the six overarching themes being explored in relation to the ages during this early period of life. This means that the chapters are subdivided by topics that are salient to that age range.

When referring to individuals of both genders, you will notice that the term 'she' and 'he' are used interchangeably. It is also worthwhile noting that the terms 'parents' and 'primary carers' are used throughout the book. This is in recognition of the fact that not all primary carers are biological parents.

You will see an annotated bibliography at the end of each chapter which will be useful for you to refer to, especially if you are particularly interested in a given topic or if you are undertaking further reading for study purposes.

Every effort has been made to provide a balanced, rational approach that supports the reader to make sense of neuroscience and its potential for application in early childhood education and care. This has meant avoiding erroneous and hence unreliable information. Instead, the focus is on providing balanced discussions throughout the text. Researching and writing this book has taken me on a professional journey, which has made me stop and think about my practice as an early childhood professional. I hope that as you read this book, you too will reflect on your own practice and may also, like me, end with many more questions than answers – but questions that will no doubt continue to drive your development.

Chapter 1
Neuroscience:
What is it?

Before we delve into the topical issues concerning early childhood and neuroscience, it is a sound idea to be clear about what neuroscience is, and its strengths and limitations. Neuroscience is the scientific study of the nervous system. Broadly speaking, the nervous system can be divided into the central nervous system (CNS) and the peripheral nervous system (PNS). In this book we are most concerned with the brain, which along with the spinal cord makes up the CNS (Bear et al. 2007). Collectively, these structures coordinate our perception of, and responses to, stimuli in the environment, enabling us to be aware of our surroundings and keep ourselves safe from danger. However, they are also responsible for our ability to think, plan, act and understand our well-being. The bundles of tissue that constitute our nervous system are at the core of how we think and feel.

Although the term neuroscience has only been in existence since the 1960s (Rose and Abi-Rached 2013; Satel and Lilienfeld 2013), evidence shows that the practice of examining the brain dates back thousands of years, with adults and children alike being treated and examined for research and medical purposes (Figure 1.1). Over time, advances in technology have resulted in techniques such as computerized tomography (CT scans), positron emission tomography (PET scans) and functional magnetic resonance imaging (fMRI), progressing our understanding of the brain as well as the identification of brain disorders and possible treatments. As a deeper understanding of the brain's anatomy has emerged, neuroscience has become an interdisciplinary field, with different subdivisions of neuroscience being developed, each focusing on the structure and function of the different brain regions and the wider nervous system. Each subdivision of neuroscience serves to advance understanding of very different areas, ranging from the organization of neurons to how different areas of the brain affect behaviour.

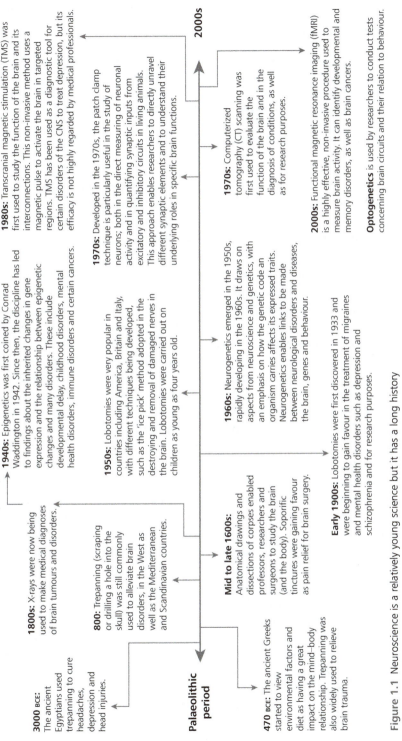

Figure 1.1 Neuroscience is a relatively young science but it has a long history

Pause for thought

1   Once you have read the time line, describe some of the key changes to neuroscience that have taken place over time.
1a Select two events from the time line that you think are particularly significant in enabling us to better understand how the brain functions.
2b Discuss why you chose these two events.

Neuroscientists can specialize in studying how different parts of the nervous system function and the optimum conditions for development – and how these can go wrong. Neuroscience is being used to further our knowledge and understanding of neurodevelopmental disorders such as **motor disorders**, foetal alcohol spectrum disorder (**FASD**), autism spectrum disorders (ASD) and **attention deficit hyperactivity disorder** (**ADHD**). In addition to investigating neurodevelopmental disorders, various techniques, including brain imaging or neuroimaging, are being used to help understand other conditions such as psychiatric disorders like **schizophrenia** and **depression**, studying healthy brains in comparison to those that display disorders.

Neuroscience enables us to begin to *see* what early childhood theorists, developmentalists and researchers have been investigating for decades. We can now start to identify the effects that early experiences have on the developing architecture of the brain – positively and negatively (Scientific Council on the Developing Child (SCDC) 2010; Shore 1997). So factors such as nutrition, health, sleep, opportunities to play, affectionate and responsive relationships and conversely, the presence of continued stress, domestic violence and chronic maltreatment are now being interpreted from brain imaging studies. When used sensibly, such findings reinforce what we already consider good practice when it comes to adopting approaches to educating very young children (Blakemore and Frith 2005). It is, however, advisable to note that the images produced from brain imaging studies do not speak for themselves and thus rely on experts to interpret their 'meaning' and implications for supporting children and their families (Rose and Abi-Rached 2013; Eliot 1999). This is crucial in preventing the misinterpretation of images derived from brain imaging studies which often result in assumptions being made. Neurologist and former classroom teacher Judy Willis informs us that 'nothing from the laboratory can be proven to work in the classroom – it can only correlate' (Judy Willis, February 2015, personal written communication).

Willis makes a point that is particularly significant in neuroscience, and indeed other sciences, that correlation (a relationship between two variables) does not imply causation. This phrase is often used when interpreting statistical data. It reminds us that just because a *correlation* may exist between two variables, it does not mean that one *causes* the other, even if it seems obvious that they are certainly correlated. Just a few examples of this include the correlation between eating breakfast and academic success for primary school children, or that watching violent television programmes causes aggressive behaviour and that gun ownership causes high murder rates. Although there may be some truth in each of these examples, there are many other variables that should also be considered and tested in determining causation. To determine causation, researchers can use controlled experiments. These often consist of two groups: one experimental group and one control group, which receives no treatment or test, thereby giving the experimental group something for comparison. Both groups are made fair by sharing as many characteristics as possible, such as sex, age, religious belief, income and education levels.

The limitations of imaging technology

Brain imaging techniques are continually developing and, as mentioned earlier, can prove highly effective in the detection and consequent treatment of certain neurological disorders. Further information concerning some the brain imaging techniques can be found in Chapter 4, but this section will outline some of the advantages and disadvantages related to a few of the commonly used techniques. The information has been included to serve as a starting point for further investigation. This is in recognition of the fact that some individuals who have little or no experience with the technology can be biased by what they see and/or hear on news media, which can often misrepresent the function, purpose and capabilities of the technology.

Table 1.1 The advantages and disadvantages of some brain imaging techniques

Brain imaging technique	Advantages	Disadvantages
Magnetic resonance imaging (MRIs)	Non-invasive. Very clear resolution images. Widely used in hospitals. Procedures are very quick, lasting a few minutes. There is no limit to the number of MRI scans one can have.	Expensive. Gives correlation but does not explain causation. Results can be inaccurate due to small bodily movements. Can cause claustrophobia for certain individuals.

Table 1.1 continued

Brain imaging technique	Advantages	Disadvantages
Functional magnetic resonance imaging (fMRI)	Non-invasive. Has led to preclinical therapeutic techniques in the treatment of certain psychiatric conditions. Does not use harmful radiation. Is widely available. Is inexpensive.	Cannot distinguish the activities of individual neurons. Has poor time-based resolution. Can cause claustrophobia for certain individuals. Requires the individual to remain completely still (can prove difficult for children and individuals with special needs).
Magnetoencephal-ography (MEG)	Non-invasive. The head can be moved during an MEG scan. Infants and children can be studied. Is particularly useful in neuroscientific research (specifically perception and processes of cognitive function). Is more accurate than an fMRI and an EEG in assessing 3-D information.	Expensive. It requires highly sensitive equipment for removing environmental magnetic interference. As yet there is inadequate evidence to support its use in the diagnosis and treatment of neurological conditions such as autism and learning disabilities.
Electroencephalogram (EEG)	Non-invasive. Identifies specific neural responses – known as event-related potentials (ERPs). Useful when studying infants as it is not sensitive to movements. Valuable in the study of language development and specific language impairment (SLI).	Does not indicate the exact brain region where the electrical activity is coming from. Weak 3-D resolution. Can detect dysfunction but not the cause. Deep parts of the brain are not well sampled.

Pause for thought

1   In your own words, write down a definition of neuroscience.
2   In which ways can it help us to understand some of the factors that affect early childhood development?
3a Do you think that some of the limitations in brain imaging could be overcome? Explain your answer.
3b In which ways might such further refinements of brain imaging techniques be useful in early childhood?

Brain structure: The basics

Know your neurons

Neurons are the building blocks of the brain. Neurogenesis – the growth of new neurons – occurs throughout much of pregnancy, slowing down only during the final trimester but continuing throughout an individual's life. Neurons have some very specific structural features that enable them to fulfil their function, as shown in Figure 1.2 – but not all neurons look like this, as this depends on their function. As the image shows, a neuron consists of three parts: the cell body and two different extensions called dendrites and axons. The dendrites bring information to the cell body, whereas the axons take information away from the cell body towards the axon terminals where connections, called synapses, form with neighbouring neurons.

Figure 1.2 The structure of a neuron

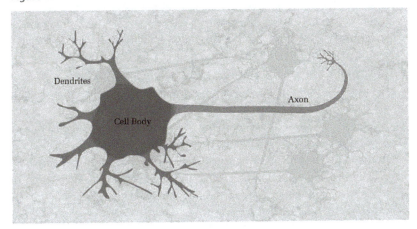

Figure 1.3 A synapse connecting two neurons

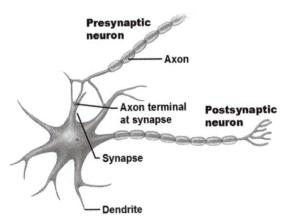

As Figure 1.3 shows, a synapse is the tiny gap between the presynaptic axon and postsynaptic area (this could be the dendrite, the cell body, or the axon of the postsynaptic neuron). Usually, the terms presynaptic and postsynaptic are used to specify two neurons that are connected. If one neuron fires (the presynaptic cell), it can chemically activate another cell on which it synapses, as shown. When a neuron is sufficiently excited, it produces an electrical signal which travels from the junction between the cell body and axon to the axon terminals. This electrical signal is often referred to as a nerve impulse but is

more correctly known as an **action potential**. Action potentials enable signals to travel very rapidly along the neuron (LeDoux 2003). Once the action potential

reaches the axon terminals it cannot travel any further because of the **synapse**, or gap, between the neighbouring neurons. However, like people, neurons need to communicate with each other, so they get round this block to the electrical signal by releasing special chemical signals called **neurotransmitters** that travel across the synapse to effect another (target) neuron. The target neuron then converts the message back to an electrical impulse to continue the process.

One thing you may have noticed and that I have not specifically mentioned is the term nerves. That's because the term actually has quite a specific meaning that is not often clear in general use. A nerve is actually a bundle of axons with a similar function; for example, the optic nerve is made up of around two million axons travelling from the retina to the brain. Axons are often covered in a fatty substance, called myelin, which acts to help conduction of electrical signals but also gives them a whitish appearance. Areas with lots of **myelination**, the term given for this white covering, are therefore called white matter, while areas lacking myelin, normally consisting of the dendrites and cell bodies, make up the grey matter. This distinction is very clearly seen in the cross-section of the brain shown in Figure 1.4.

Figure 1.4 White and grey matter in the brain

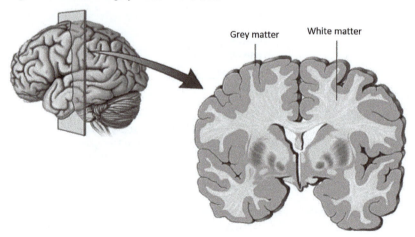

The description above uses only a single neuron, communicating with a single target neuron, but you may have spotted that the optic nerve contained two million axons. At birth, the brain contains approximately 86 billion neurons (Azevedo et al. 2009; Herculano-Houzel 2009)! But by no means is the brain fully developed at this early stage of life. In addition, remember this is the number of neurons and each neuron will form many connections with target neurons. As a general rule, 86 trillion bits of information move around your brain every second! Take a look at the numbers below.

Number crunching

Here's the maths (Azevedo et al. 2009; Herculano-Houzel 2009):
86 billion neurons
x
10 action potentials per second (2 Hz to 200 Hz range)
x
100 synapses for each axon (estimate)
=
86,000,000,000,000 bits of information transmitted per second!

Mirror neurons

Mirror neurons are neurons in the brain which are activated both when actions are executed and the actions are observed, playing an integral role in deciphering other people's future intentions (Acharya and Shukla 2012). So, they might become activated when an individual observes another carrying out a movement or when observing another experiencing an emotion such as happiness, fear or pain. For example, an adult might tell a child they 'feel their pain' if they have just had a fall. One definition is provided by Rose and Abi-Rached (2013: 145), who explain that mirror neurons 'fire' an action potential:

> When an individual observes a movement being carried out by another –
> usually a 'conspecific' but sometimes an individual of another species – a
> small number of neurons are activated in those areas of the brain that are
> also activated when the individual carries out the same movement him- or
> herself.

Mirror neurons were accidentally discovered by Rizzolatti and Gallese in 1988 in the frontal cortex of macaque monkeys. Since then, the role of mirror neurons in human development – particularly in empathy – has garnered both positive and negative attention from neuroscientists. (This will be discussed a little later in this section.)

Mirror neurons might play an important role in early childhood develop–ment – especially with regard to emotional and social development (Hamilton and Marsh 2013; Ramachandran 2010; Gazzola and Keysers 2009; Rizzolatti and Fabbri-Dastro 2008; Frith 2007; Decety and Jackson 2004; Gallese 2001). Consider the life of a newborn baby – its primary methods of communication are dependent on eye contact, gestures and vocalizations. From birth, an infant is primed for socialization; this might manifest when an infant imitates their caregiver when they poke their tongue out at them, or smiling back at their caregiver. This is depicted in Figure 1.5.

Figure 1.5 Mirror neuron system in action

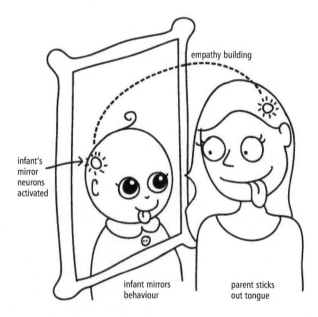

infant's mirror neurons activated

empathy building

infant mirrors behaviour

parent sticks out tongue

As the infant gets a little older, it is able to observe the adult's facial cues and use the information to interpret whether situations are safe or pose a threat to their well-being. This social referencing is made possible when the adult and infant share a healthy emotional bond, one in which the infant feels safe and secure and can predict their caregiver's responses to them. If, on the other hand, the bond between the infant and the adult (for example, their parent or primary carer) is unstable and weak, the infant might find it difficult to interpret others' expressions and find it difficult to express themselves through fear of being reprimanded or not responded to positively (Twardosz and Lutzker 2010; Laible 2004; Denham and Kochanoff 2002). In his paper entitled 'Embodied simulation: From neurons to phenomenal experience', Gallese (2005: 36) explains:

> Emotions constitute one of the earliest ways available to the individual to acquire knowledge about its situation, thus enabling her/him to reorganize this knowledge on the basis of the outcome of the relations entertained with others. This points to a strong interaction between emotion and action.

Mirror neurons can also shed light on the impact of autism, which typically affects a child's emotional and social development, with resulting difficulty in identifying and interpreting other people's feelings. Research conducted by Dapretto et al. (2006) found that the mirror neuron system of children with autism showed very little or no activity compared to peers who did not have autism. This inactivity of the mirror neuron system is compounded by the fact that some children with autism tend to avoid eye contact, which means that their mirror neuron system does not get the training it needs to improve function. Iacoboni (2012: 167) speculates that:

> If we understand this incredible system better, it will in the long run provide obvious opportunities to train empathy and thus social competence.
> Imagine how much you could achieve by using this in a school context.

However, mirror neurons are a controversial subject among some neuroscientists. The main cause of this controversy is the concept of *action understanding* (the notion of the mirror neurons mirroring the behaviour of the other, as though the observer were itself acting). Rizzolatti et al. (2001: 661) identify that the popularity of mirror neurons theory is also partially due to the ease with which we can relate to it:

> We understand action because the motor representation of that action is activated in our brain.

The general overzealous attitude to mirror neuron theory has led to its indiscriminate implementation concerning action understanding, which has served to prevent its acceptance as a reliable, tested theory that is directly applicable to humans. Hickock (2009: 5) puts forward one reason for the uncertainty of the efficacy of mirror neurons:

> The intense focus on one interpretation of mirror neuron function, that of action understanding, has impeded progress on mirror neuron research. Although the action understanding hypothesis is interesting and worthy of investigation, I will argue that it fails dramatically on empirical examination.

Hickock (2009) makes a valid point – for a theory to be accepted as true, it needs to be observable and, in the case of mirror neurons, directly applicable to humans. Yet a central problem in the case of mirror neurons is that it has been generalized to humans without systematic validation. Another principal concern is that action understanding has been shown to be achieved via non-mirror neuron mechanisms as well as the mirror neuron system (Rizzolatti and Craighero 2004; Rizzolatti et al. 2001), which means that their role in the development of socio-emotional skills may have been overestimated. The take-home message, therefore, is to exercise caution when reading about the wonders of mirror neuron theory. It is one way of comprehending action understanding and, to an extent, empathy – but it certainly is not the only way.

Pause for thought

1   What do you think is the role of mirror neurons in early emotional and social development?
2   What are the implications for parents and practitioners, particularly during their interactions with babies and children?
3   Do you think that mirror neurons contribute to communication and empathy? Explain the reasons for your answer.

The effects of neglect

You have just read about mirror neurons and hopefully you will have noticed that there is quite a lot of research investigating these neurons. Some of this research has shown positive results concerning their role in socio-emotional skills, while others show that they are not as significant as we thought. Consequently, there is no clear consensus on their function in humans. However, it is not always the case that misunderstanding and over-extrapolation arise from a lack of consensus; sometimes they simply arise from an individual study. Look at the image shown in Figure 1.6 depicting the effects of neglect on the brain. This now infamous image depicts brain scans of two contrasting three-year-old children's brains (in which the 'normal' brain is significantly larger than the brain of the child who has suffered extreme neglect). The image of the brain depicted on the right is taken from a child who had suffered extreme institutional abuse in an orphanage in Eastern Europe and therefore has virtually no predictive value in a different context. That said, it is easy to understand why non-experts or the layperson could be seduced by this image, due to the visual impact of the substantial difference in size between the two brains. Such misinterpretation

Figure 1.6 Bruce Perry's impact of neglect on brain development

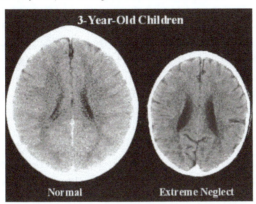

and over-extrapolatation of images from brain studies like Perry's is characteristic of neuroscience being used inaccurately. This image illustrates the importance of giving practitioners access to evidence that is reliable, replicable and appropriate for their field.

The importance and meaning of Perry's work for more typical development has been overexaggerated through misinterpretation and over-extrapolation, but it has not led to any widely-held and yet false beliefs about the brain, so-called neuromyths, which are often used to promote a particular brand or way of doing something. Further on in the chapter are a few examples of neuromyths which pervade current thinking. The examples are not exhaustive, as I have selected just a few that I think are most relevant to early childhood. Those wishing to find out more should refer to the reading list at the end of this chapter.

The developing brain

Now that you have a basic overview of the building blocks of the brain, let's start to think about development specifically. I mentioned above that at birth the brain is not yet fully developed, so what exactly happens to the brain as we go through life? The answer to this question is very broad and you can find whole books on this topic but this chapter focuses on a few specific issues that are now well understood. This understanding comes from research studies – some with animals and some with humans. Some of the research is conducted post-mortem by analysing tissue and some is conducted in the awake person using modern imaging techniques. Through these converging methods there are now some well-substantiated explanations of brain growth and development. Let's start with the synapse.

Synapses and development

At birth, each neuron in the cerebral cortex has approximately 2,500 synapses. By the time an infant is three years old, the number of synapses is approximately 15,000 per neuron. This amount is roughly twice that of the average adult brain (Rogers 2011; Pettus 2006). Plasticity is therefore at its most rapid during early childhood, when the brain is most sensitive to environmental influences. Neural connections grow and are strengthened in response to these experiences, be they positive or negative. Repetition of experiences leads to neurons creating pathways in different parts of the brain, based on the experiences. Perry (2001) tells us that experience, good and bad, literally becomes the **neuroarcheology** of the individual's brain.

I mentioned that each neuron forms many synapses with other neurons, and the process of synapse formation, or **synaptogenesis** as it is correctly known, is highly prolific during the first five years following birth. Figure 1.7 shows this rapid increase in synaptic connections between birth and adulthood – with few synaptic connections at birth, swiftly increasing within months and by two years closely resembling that of an 'average' adult's brain. This indicates that synaptic growth is greatly dependent on environmental input, occurring directly as a result of the experiences and interactions that the baby is part of. This can be through play, exploration, communicating with others and enjoying secure and responsive relationships, with opportunities for repetition for mastery of skills which strengthen the existing synaptic connections (Goswami 2015). These connections will continually change throughout a lifetime to adapt to new challenges and cognitive aging. So although you may have a similar number of synapses throughout aging (which decreases as we get older), these are not the exact same set of synapses that you have during your lifetime.

Figure 1.7 Quantity of synapses at birth into adulthood

Newborn   1 Month   9 Months   2 Years   Adult

Synaptogenesis occurs at a far greater rate in early childhood than in adulthood, making the early years of a child's life a highly significant period in terms of learning and development. The significance of this is highlighted by Linderkamp et al. (2009: 11), who explain their significance:

> After the first year of postnatal life the total synapse number slowly increases and reaches the maximum at five years when the child's brain weighs almost as much as in adults. During the first five to ten years of life, the child achieves the highest number of synapses, thereby enabling the child to acquire enormous behavioural social, environmental, linguistic and cultural information.

Synaptic pruning and neural growth – use it or lose it!

Pruning refers to the elimination of synaptic connections over the course of human development, specifically shortly after birth and during adolescence. Pruning is critical to brain growth and learning. The process of pruning can be likened to the regular pruning and cutting of flowers and trees – as a result of the pruning process, existing structures are strengthened which ultimately promote healthy growth. This rule can be extended to the developing brain, for which pruning is necessary in order to remove the connections that are not used, because as the infant's understanding of how the world works becomes more complex as she grows, unused connections are no longer needed and hence 'die'. Pruning thus strengthens existing connections in the brain, through the infant's exposure to experiences and interactions within their environment. Pruning also makes room for new connections that are made between neurons as the child learns and masters the detail of evolving concepts. So, it is not a matter of how many synapses you make a day, but rather the specificity of these synapses that allow for learning to occur.

Synapses change in efficiency with each experience (LeDoux 2003), with unused connections dying away and making room for new growth, with new information coming in from the senses, changing the landscape of the brain. This process can be linked to Piaget's concept of equilibrium and disequilibrium (1964). According to Piaget, when knowledge is easily assimilated, the child is in a state of equilibrium (balance); however, when their existing knowledge needs to be accommodated (changed), the child enters a states of disequilibrium (imbalance). This unsteady state continues until the child can modify her existing schemas (or representations) to take in the new information. Hence, a state of disequilibrium drives the process of learning as young children do not like to be puzzled by new information which they cannot fit into existing schemas (Athey 1990). As the child has more experiences, new information is used to modify or add to existing schemas. Consider a two-year-old child who, as a result of her experiences, understands that all dogs are furry, four-legged animals. When she first sees a cat, she may well say 'Look, dog! Woof!' If her mother corrects her and says 'No, that's not a dog; it's a cat,' the child's existing knowledge is adapted to take in the new information. The diagram below summarizes the process of equilibrium and disequilibrium.

Figure 1.8 Equilibrium and disequilibrium – making room for new information

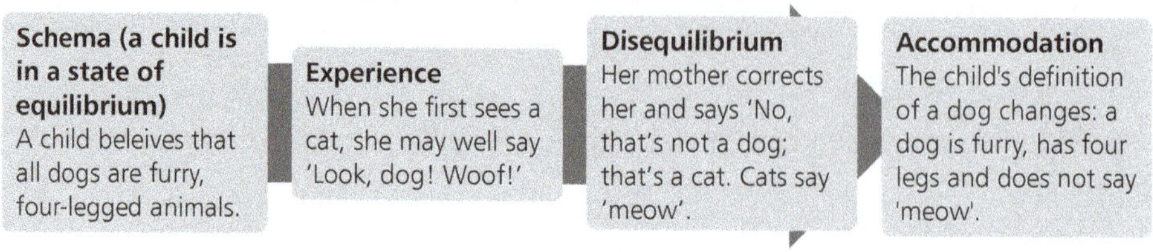

Schema (a child is in a state of equilibrium) A child beleives that all dogs are furry, four-legged animals.	Experience When she first sees a cat, she may well say 'Look, dog! Woof!'	Disequilibrium Her mother corrects her and says 'No, that's not a dog; that's a cat. Cats say 'meow'.	Accommodation The child's definition of a dog changes: a dog is furry, has four legs and does not say 'meow'.

Repetition of experiences as part of successful learning is therefore vital in relation to specificity of connections and plasticity. Having the opportunity to regularly practice skills such as language, movement and exploration and problem-solving during particular activities enable the infant to build familiarity, confidence and competence. For this reason, repetition is considered to be an essential influence in promoting neural growth and learning. The brain's ability to constantly adapt through learning facilitates its lifelong plasticity (see Chapter 2). Plasticity refers to the brain's unique ability to change its structure and function as a result of changes internally (within the body) or externally (in the environment). Kolb (2009: 1) says:

When we first learn a new motor skill, it seems impossible until practice – repetition – changes the brain.

Pause for thought

1 Explain to your colleague or a fellow student how healthy brain development is dependent on both the growth and pruning of synaptic connections.
2 What effect does repetition of activities have on synaptic growth?
3 How could you incorporate your knowledge about synaptic growth into your setting's planning of activities on a regular basis?

Sensitive periods

The concept of sensitive periods refers to distinct phases during early childhood when the brain is best able to receive and use information gained from experience, in order to learn specific skills. The period of birth to five specifically represents a sensitive period for babies and children as it represents a time of fervent growth and development, with neural connectivity being at its most prolific (Chugani 1998). Learning specific skills still occurs beyond this period, but the brain does become less malleable as the neural networks get laid down and consolidated over time and with repetition of multi-sensorial experiences. Thomas and Knowland (2009: 2) inform us that sensitive periods are understood by the speed of change which occurs in behaviour at different ages:

> A sensitive period represents a window within which the effects of environmental stimulation on brain structure and function are maximised. The main source of empirical evidence that informs our understanding of sensitive periods is the rate and upper limit of behavioural change that individuals can achieve at different ages.

It is thus a crucial time for acquiring motor skills, language and forming attachments. If the necessary information is not received during this sensitive period, the pathway in the developing brain might not acquire the ability to process information and, as a result, perception or behaviour can be undermined. Take motor skills – babies and young children learn through the movement and coordination of their bodies. This movement not only strengthens muscles but also boosts brain development; controlling body movements leads to control of finer movements such as being able to manipulate and explore play materials and, later, learning how write. It is difficult to master such fine motor skills if control over larger movements of the body is poor. If all young children are to successfully learn during sensitive periods, it makes sense that the concept of sensitive periods should be reflected across the early childhood curriculum provided in the setting. Such consideration to planning of the physical environment and the resources in it can make a positive difference to young children's ability to acquire skills and consequently build confidence. That's not to say that practitioners must fit in as much teaching and learning as possible as early as possible, but to pay close attention to the layout of the environment and quality of multi-sensorial experiences provided. *Are they geared to the development of the individual child?* To conclude, the concept of sensitive periods is certainly not part of the neuromyths that so commonly pervade the discourse concerning educational neuroscience. Thomas and Knowland (2009: 4) leave us with something positive to ponder with regard to the practical application of understanding sensitive periods in children's development and learning:

> Exciting vistas for the future include the possibility of using genetic and brain-imaging data to identify the best developmental times for training new skills in individual children.

When practitioners understand the crucial role of sensitive periods in facilitating early brain development, they can think more critically about their planning of the environment and what changes can be made to maximize each child's learning experience. Of course, it is not only knowledge concerning sensitive periods that enables this, but it can certainly add to practitioners' existing knowledge base. Given that this tends to be drawn from cognitive theorists such as Piaget, Bruner and Vygotsky, as well as Bowlby's theory of attachment, combining this understanding with more current, scientific knowledge can further improve the quality of early education and care.

Pause for thought

1 If a child continually misses out on a specific learning opportunity during the sensitive period, her brain may not develop its circuitry to its full potential for that specific function. Does this mean that cognitive development will be impaired? Explain the reasons for your answer.
2 Consider the concept of sensitive periods for learning; is its significance reflected in the planning of experiences in your setting?
3 How far can knowledge concerning sensitive periods influence curriculum planning for the birth to threes?

Myelination

I mentioned above a process called myelination, where the axons of neurons become covered in a white fatty substance that helps conduction of the action potential. Like changes in synapses, myelination is an important developmental process and one that continues into our twenties (Fields 2008). This prolonged period of myelination indicates that, like synaptic changes, it might be experience-dependent. Neuroscientists now have imaging techniques that allow them to examine myelin and have indeed found that it can change in responses to environmental experiences. Much of the direct evidence comes from work in animals but there is also correlative evidence from people, showing myelination correlates with learning, development of skills and memory (Fields 2008). Indeed, one worthy study shows that the amount of myelination in a structure increased proportionately to the number of hours a person had practised playing a musical instrument (Bengtsson et al. 2005).

As part of their study, these researchers used diffusion tensor imaging to measure myelination on a group of eight (age-matched) male musicians, specifically demonstrating the importance of practising in early childhood in terms of white matter plasticity – compared to its limited plasticity in adulthood. For further information concerning this study, please refer to the bibliography.

Neuroscience: Distinguishing between the neuromyths and the facts

The proliferation of scientific information available since the 1990s, which was proclaimed as 'the Decade of the Brain' by the US Congress (Goldstein 1994), has resulted in exciting developments in the field of neuroscience. However, it has also left the field open to misinterpretation and over-extrapolation. In this section I am going to look at a few examples of this.

Brain-based programmes

Efforts continue to be made in bridging the gap between neuroscience and education, and some ground-breaking work has been carried out in this area (Oates et al. 2012; Howard-Jones et al. 2007; Shonkoff and Phillips 2000). However, there still exists an overabundance of unfounded claims supposedly supported by neuroscience, particularly in education. Some of these claims have been used to develop commercial products in the area of brain-based learning programmes. One such product is Brain Gym. Brain Gym is designed to boost learners' performance in the nursery and classroom and was widely used across schools in England until the late 1990s. According to the product, learners' performance can be improved in the classroom through a set of twenty-six exercises aimed at promoting hand-eye coordination, flexibility and listening skills. Before we look more carefully at this claim, let's have a look at one of the exercises.

Brain Gym exercise

Below is one set of exercises that is used as part of the Brain Gym approach to learning. It is included for two reasons:

1 To alert you to the claim that 'water helps the body eliminate stress and focus'.

2 To share some of the students' feedback concerning this set of exercises, which they undertook to ascertain their thoughts on its effectiveness in promoting learning.

Read the information in the box below and have a go at this by yourself before answering the questions underneath.

How to use Brain Gym

1 Drink Water. Water helps the body eliminate stress and focus.
2 Press lightly with your thumb and index finger on the indentations that are just below your collar bones. Let your other hand rest gently across your belly button. This is called activating your 'Brain Buttons' and helps with focus.
3 Stand up and place the right hand on the left knee while lifting the knee to the hand. Do the identical movement for the left hand on the right knee, as if marching in place. This is called 'cross crawl' and integrates the movements on both sides of the brain.
4 Stand up and cross the right leg over the left at the ankle. Put the right wrist over the top of the left one and intertwine the fingers. Turn the fingers in towards the body and rest them on the centre of the chest. Breathe deeply. This is called 'hook-ups' and helps energize and centre the body.

Pause for thought

1 How did the exercise make you feel?
2 Did it improve your ability to focus?
3 Do you think the exercise is effective in helping children to learn? Explain the reason for your answer.

Feedback from learners

At the start of one lesson with learners on a foundation degree in early childhood studies (which consisted of nursery managers and practitioners), Brain Gym was introduced with an explanation of how and why it has been used across a number of nurseries and schools. Having completed the exercises, the learners then evaluated the effectiveness at the end of their lesson. All learners said that the set of exercises made no difference to their ability to focus or to learn during the lesson. They also said that they would not adopt it as a resource in their respective settings, mentioning that the exercises were actually a little difficult to follow and carry out. One nursery manager said:

> I was too busy trying not to fall over – maybe that's why it claims to improve focus!

Learners did identify the benefit of drinking water and exercising regularly while stating that 'it's not brain science to know that we have to keep our children healthy so that they are better able to learn'. Early childhood educators would do well to follow up claims made by brain-based educators and any references contained within their curriculum. This includes assertions made about being *brain-based* and *derived from scientific research*. Learners on this foundation degree are taught about research methodologies as part of a module on action research, so they understand what scientific research involves and how to check for reliability. This can help early childhood practitioners feel more secure in the knowledge that the information presented is from well-founded sources and thus more applicable in their daily work with babies and children.

The evidence for Brain Gym

The experience of the learners was not in support of Brain Gym as a brain-based technique, but Brain Gym is supported by some classroom observations and teachers' reports within the programmes which are designed to improve young children's concentration, memory, problem-solving ability and processing of information. However, irrespective of reports from individual teachers and learners, it is important to look at the scientific evidence for the claims made by the product.

Let us start with the claim about the benefits of water regarding cognitive, emotional and mental health. We understand the importance of drinking enough to keep hydrated and its role in the regular functioning of our cells and organs. However, to assert that it 'eliminates stress' is false and misleading: it is not a magical antidote for stress (i.e. drinking water cannot reduce the amount of the stress hormone cortisol, released in times of heightened anxiety!). The concept of drinking water to improve learners' focus is also erroneous; unless the child is dehydrated, drinking a glass of water will not cause an improved ability to learn (Rogers et al. 2001).

But what about the specific exercises – could they be beneficial? Well, there is some evidence that aerobic exercise may be beneficial per se, and Brain Gym could be tapping into this (Howard-Jones 2007), but this is not based on neuroscientific evidence, which renders some of the claims unreliable. So, to base an entire teaching philosophy such as this, on neuroscience has proven to be misleading. Howard-Jones et al. (2007: 8) inform us that:

> Strategies, such as those found in brain gym, have not been scientifically or educationally assessed with any rigour, but often use pseudo-scientific explanations to support their credibility.

Ben Goldacre is well known for his blogs, newspaper column and books that debunk the pseudoscience underpinning brain-based programmes, including Brain Gym. When discussing the exercise included here, he quips:

> If only the children could do the hook-up exercises, this would connect the electrical circuits in the body, containing and thus focusing both attention and disorganised energy, and they would finally see sense. (Goldacre 2009: 14–15)

In summary, the claims made by the founders of Brain Gym that doing the exercises promote learning are unsupported by peer reviewed research into either the outcome (i.e. improved learning) or the mechanisms (altering activity in the brain in a specific way) (Goldacre 2009; Howard-Jones 2007), resulting in Brain Gym being largely discredited as a trustworthy teaching resource that promotes readiness to learn. The dangers of such misleading pseudoscience occur when teachers and early childhood practitioners take this information as unquestionable, applying them with children in nurseries and schools, without assessing the reliability of the claims put forward. Research carried out by Howard-Jones et al. (2007) established that teachers appreciate greater access to evaluative evidence that examines their scientific origin and their effectiveness.

Critical periods

The concept of critical periods refers to a specific time frame within which a child 'must' acquire certain skills, because if this does not happen the child cannot easily attain the skills outside this time frame. Critical periods broadly cover the birth-to-three age range, but different skills are acquired at different points within this range (Bruer 2011). The concept has swung from being regarded as one of the key tenets of neuroscience to being deemed erroneous – another myth that permeates through early childhood policies

(Bruer 2011; Field 2010; Marmot Review 2010; Wilkinson and Pickett 2010; Howard-Jones et al. 2007; Thomson 2001; Chugani 1998). Although alternative concepts have followed, such as 'sensitive periods' and 'windows of opportunity', these are not used in a measured way, but interchangeably depending on the publication. It is therefore prudent to avoid such restrictive labels and instead use your judgement based on the children with whom you work and the behaviours shown, when considering the type and duration of support that may be required to best nurture the child's development.

The concept of sensitive periods is a more useful way to understand brain development, as individuals, generally speaking, have the capacity to learn and develop throughout their lives – brain development does not end rigidly at three years, allowing for no more growth and learning after this. If this were the case, intervention and individualized support programmes for those children who have had impoverished and deprived early experiences would be a waste of effort and time.

Case study

Ahmed was placed in foster care at the age of seven due to experiencing ongoing neglect and maltreatment from his mother and her partner. Emotionally and socially he was severely behind expected developmental norms for his age. He used to collect rubbish off the streets to play with as toys and had to be taught how to use cutlery. He could not demonstrate affection and experienced intense difficulty with communicating, not knowing how to build friendships. Three years have passed and he is doing very well. We manage difficult and demanding situations together with Ahmed and give him lots of encouragement, patience and love. It's really important not to be judgemental because so much of his behaviour was his way of trying to deal with what was happening to him. He demonstrates affection, can accept affection, and is slowly developing compassion and empathy. Ahmed's communication skills have vastly improved and he feels like a member of the family. It has been a slow and very painful journey but well worth it.

Pause for thought

1 Do you think that the concept of critical or sensitive periods is relevant in the discussion concerning early childhood experiences and brain development? Explain the reasons for your answer.
2 To what extent do you think the attachment relationship is important in healthy brain development?

Advertising of food and drinks

Drinks such as Neuro (sold in the US and previously in the UK), which advertisers claim promote alertness and relieve fatigue (to name but two 'benefits' of the drink) prove very popular, particularly among students who wish to boost their academic performance. However, there are no scientific studies which support the claims that this *scientifically-based functional drink* directly benefits individuals' academic performance or ability to concentrate. Ultimately, this does not seem to matter much given that the language and imagery used in such advertising campaigns can easily lure those in need of enhancing their performance to buy the product. Yet, if you read between the bold claims, the advertisers do state in very small writing that:

> These statements have not been approved by the Food and Drug Administration.

Apart from adopting age-old strategies such as undertaking regular revision and adopting learning strategies that *do* directly impact on the brain's ability to recall important information, consumers really should get into the habit of paying closer attention to food and drink labels and do a little research into claims put forward by products.

Pause for thought

1 Reflect on the some of the neuromyths explored in this chapter. How might you respond to a colleague who says 'The science doesn't matter – the children in my class do the exercises and they focus better as a result!'
2 Discuss two ways in which neuromyths may ultimately work against educational achievement.

Neuromyth or fact?

Have a go at deciding whether the statements below are true or false.

		True	False
1	Fundamental pathways are present in the brain before birth.		
2	Babies are born with the ability to learn all the languages in the world.		
3	A human baby's brain has the greatest density of synapses by the age of three years.		
4	The first three years of life can be the most critical for brain development.		
5	Good nutrition is one effective way to aid healthy brain development.		
6	Reading to a newborn infant is the best way to help a child learn to read in the future.		
7	There are times when a negative experience or the absence of appropriate stimulation is more likely to have serious and sustained effects on a child.		
8	The large majority of what we have learned about the brain comes from research conducted on animals.		
9	Brain research has been misunderstood and misapplied in some contexts.		

The answers to this quiz can be found at the end of this chapter.

How can we wise up to the neuromyths?

A key reason for the persistence of neuromyths is the absence of a suitable forum in which professionals from neuroscience, psychology, early childhood education and care come together. Where such forums exist, it can still be challenging and time-consuming for early childhood practitioners, advocates, parents and primary carers to investigate the trustworthiness of the sources of information they are presented with. Individuals may also not know how to do this or, indeed, what to look out for. Issues surrounding different research methods in neuroscience, including the questions of who interprets these and for what purpose, can leave individuals confused. In relation to this, Shonkoff and Bales (2011: 11) identify three challenges in translating the science behind early childhood development:

- Determining what needs to be translated.
- Identifying obstacles to public understanding.
- Developing and verifying the impact of specific frame elements that improve public thinking.

Further to the challenges identified, consideration needs to be given to *who* determines what needs to be translated, *how* the evidence is disseminated to the public and *what* aspects of the evidence are disseminated. Such complex issues clearly need to be resolved through closer collaboration between professionals in early childhood education and neuroscience, so as to ensure that the challenges and opportunities arising from both fields are equally represented and examined (Wolfe 2010; Teaching and Learning Research Programme 2007). This can result in a significant difference in how early childhood professionals understand and interpret neuroscience, and similarly, neuroscientists might be able to develop a clearer, practical understanding of how neuroscience can inform early childhood policy and practice. Howard-Jones et al. (2007: 6) explain:

> What is clear is that it is important for educationalists and teachers along with scientists and researchers to share together what they are finding out about successful learning in this new interdisciplinary field of neuroscience and education.

How we can separate the myths from the facts

- Unequivocal teaching/training that includes identification of current pervading neuromyths and teaching of experimental design to enable teachers to evaluate research for themselves.
- Introducing neuroscience and early childhood education books as mandatory reading on early childhood studies courses. Ideally, this should be done through collaboration between the two fields.
- Increased accessibility to online articles and journals free of charge.
- Reading blogs written by neuroscientists to keep up to date with developments in neuroscience and education.
- Exercising due caution when it comes to adopting brain-based programmes – follow up the supporting evidence that is listed.

Concluding thoughts

This chapter has attempted to give a useable definition of neuroscience, together with explanations of some of the well-understood changes to the brain that take place in early childhood. You have also learned, through discussion of various examples, about the issues commonly associated with neuroscience, with emphasis on the myriad of myths that pervade the field and that of early childhood education, with practical suggestions given throughout the chapter on what you can do to separate the myths from the facts – and how some of the more reliable information might be used to strengthen good early childhood practice.

Further reading

Howard-Jones, P., Pickering, S. and Diack, A. (2007). *Perceptions of the Role of Neuroscience in Education.* London: The Innovation Unit.

This publication provides a brief historical overview of neuroscience and explores some of the common myths surrounding neuroscience and its application in early childhood classrooms. Detailed analysis of teachers' views regarding the use of neuroscience in education is also included, as well as some of the present issues and potential solutions as to how these might be overcome. A must-read for teachers, trainers and early childhood practitioners who want to learn more about the relationship between neuroscience and early childhood education, and how some of the answers provided can be put into action.

Oates, J., Karmiloff-Smith, A. and Johnson, M. H. (2012). *Early Childhood in Focus 7: Developing Brains.* Milton Keynes: Open University.

This publication is divided into three broad sections, which examine prenatal brain growth, early brain development and the wide range of factors and environmental influences that affect this complex process. The policy questions included at the end of each section help to contextualize the issues discussed in relation to early childhood practice. This is an excellent, comprehensive resource for early childhood practitioners, researchers or primary carers wishing to gain clear insight to the brain and their role in helping to shape optimal brain development.

Sigman, M., Peña, M., Goldin, A. P. and Ribeiro, S. (2014).

'Neuroscience and Education: Prime Time to Build the Bridge'. *Nature Neuroscience* 17 (4): 497–502.

As the title infers, the main message of this thought-provoking paper is that the time has come to start adopting neuroscience-based theories to inform educational practice with young children. The researchers examine four cases that demonstrate how neuroscience interacts with other disciplines to support education and different aspects of human physiology (including nutrition, exercise and sleep). The use of neuroscience in detecting cognitive irregularities in infancy is examined, with references to neuroscience methods and theoretical frameworks to help build understanding of how children acquire language and learn how to read.

Neuromyth or fact? Answers

1 Fundamental pathways are present in the brain before birth. **True**.

2 Babies are born with the ability to learn all the languages in the world. **True**.

3 A human baby's brain has the greatest density of synapses by the age of three years. **False**.

4 The first three years of life can be the most critical for brain development. **False**.

5 Good nutrition is one effective way to aid healthy brain development. **True**.

6 Reading to a newborn infant is the best way to help a child learn to read in the future. **False**.

7 There are times when a negative experience or the absence of appropriate stimulation is more likely to have serious and sustained effects on a child. **True**.

8 The large majority of what we have learned about the brain comes from research conducted on animals. **True**.

9 Brain research has been misunderstood and misapplied in some contexts. **True**.

Chapter 2
Why Should Early Childhood Practitioners Know About Neuroscience?

What to expect from this chapter

This chapter presents some of the factors which currently impede opportunities to incorporate knowledge and experience from neuroscience in early childhood education and care. Discussions will be included regarding how neuroscience is helping to advance professionals' understanding about the developing brain and how this knowledge could be used to inform early childhood education policy and practice. Evidence-based research that has emerged from the collaboration between neuroscientists and educationalists will also be examined, with questions to encourage reflection about their significance. Thought will be given to the existing gap between early childhood provision and neuroscience, alongside some of the factors that have contributed to this 'divide' and recommendations to help bridge this gap.

Early childhood and neuroscience: Does a gap exist?

Like most disciplines, neuroscience has taken decades to be readily accepted as a science which can reliably contribute to influencing education – particularly early childhood education (Bruer 2011). Issues of research ethics concerning the use of brain imaging techniques with very young children and the over-extrapolation of findings loosely based on neuroscience have not only led to this gap but also resulted in apprehension in grappling with knowledge that is neuroscience-based (Ansari et al. 2011; Devonshire and Dommett 2010; Goswami 2006; Jolles et al. 2005). While attempts have been made to use some evidence derived from neuroscience in education, these have generally been erroneous.

While conducting a search concerning the application of neuroscience-based evidence in early childhood education, one issue quickly emerged: national and international studies continue to capture the professional experience and views of primary and secondary educators, but no evidence (as yet) exists which elicits the thoughts and opinions of professionals from the early childhood sector about the relevance of neuroscience to education (Dekker et al. 2012; Howard-Jones et al. 2007; 2009). Herein, lies a key factor as to why early childhood professionals and neuroscientists are yet to exchange ideas about the use and evaluation of findings from neuroscience to inform early childhood provision. Without this type of collaboration, we have little scope for cultivating early childhood programmes in light of evidence derived from neuroscience.

During a conversation with one of the leading researchers in the studies cited above, he was asked why this is the case. The response confirmed the assumption that the views of early childhood practitioners were not sought in any of the studies and that 'this is a conversation that should happen, in order to explore what neuroscience can contribute to practice and how' (Howard-Jones, March 2015, personal written communication). Another significant barrier is the media coverage concerning neuroscience. While we can benefit from articles that highlight the consequences of misusing neuroscience in education, some articles can be hyper-critical and almost dismissive of key theories that are not only the cornerstone of early childhood practice but also inform early childhood policies (Gerhardt 2015; Williams 2014). Such scepticism and suspicion serves to further distance contemporary early childhood issues – and professionals – from neuroscience.

Neuroscience does offer insights concerning the young brain and the vital role that adults play in helping to shape it, both in the short and long term – which can be applied to practice (Spitzer 2012; Howard-Jones 2010; OECD 2007). What does undermine its value in early childhood education and care is the lack of understanding and inaccuracy in which it is sometimes presented – which is then exacerbated by enthusiastic but misinformed practitioners who consequently disseminate erroneous information.

Pause for thought

1   Do you or your colleagues perceive a gap between neuroscience and early childhood practice?
2   What reasons are given for this gap?
3   In which ways could a brain-based approach improve your current practice?
4   Discuss two practical ways you could inform and inspire your colleagues about adopting a brain-based approach to your work with young children.

What are the implications of neuromyths on early childhood education and care?

At present, we are left at an impasse because if this gap did not exist, or was not so bad, then the majority of the myths would not have the opportunity to thrive or persist as they do. As a result of such pseudo-research, early childhood practitioners and teachers alike might readily accept supposed findings and hence adopt teaching methods which are informed by questionable research that does not stand up to its claims. Recent research undertaken by Howard-Jones et al. (2007: 4) concluded that:

> Current teacher training programmes generally omit the science of how we learn, so the information that teachers are getting comes from a number of sources which includes the general media, conferences, in-service training courses, books, materials and journals.

Basing practice and general teaching and learning methods on research that is not evidence-based can impede progress. For example, the possibility of evaluating and further refining research and practice can be hindered due to the research informing it being of indeterminate origin. Ultimately, all this leaves neuroscientists frustrated at the neuromyths which abound, making them less amenable to bridging the gap between neuroscience and early childhood education and care. Another possible consequence can be inconsistent practice, which varies in the types of opportunities afforded to babies and young children. This might be a result of some settings and professionals trying to interpret and apply findings from neuroscience while other settings do not engage with it at all.

Pause for thought

1 Suppose a parent of one of your key children asks you to explain why educators are interested in the brain. Make a few notes about how you would respond to this parent.
2 Explain the value of professional development which is informed by neuroscience.
3 Why do you think that some of the research on early brain development is often misinterpreted by early childhood advocates?

How might this gap be bridged?

Neuroscientists and educational researchers are undoubtedly producing worthwhile evidence concerning the integration of neuroscience in education, but we now need a similar approach to early childhood education and care. This would be timely given the raft of policies and attempts at intervention that have come under scrutiny for their lack of rigorous scientific testing and evaluation (Allen 2011; Field 2010; Allen and Duncan Smith 2008).

The good news is that neuroscience *is* being employed by diverse (and in some cases overlapping) professions to help reinforce and enhance their practice (Evans 2015; Cozolino 2013; Royal College of Midwives 2012; Immordino-Yang and Damasio 2007). One example (which will be discussed in greater depth in Chapter 3) is provided by a trauma parenting specialist and trainer who uses neuroscience with the children, young adults and families as part of the therapy provided. She explains:

> I use neuroscience to gain an insight into how the brain operates based on early childhood experiences. This offers them an opportunity to use this neuroscience to start to work with their highly reactive brain and the fear/threat cycle in it which otherwise dominates and wreaks havoc in their daily lives and relationships. (Evans 2014, personal written communication)

This is just one example of how professionals working with children and their families understands and *translates* neuroscience into her daily work. Sharing good practice and working in a multidisciplinary manner can help practitioners to broaden their knowledge about their role in supporting early brain development. Embedding neuroscience at a policy level can also help to ensure that evidence obtained from neuroscience can be filtered down to practice across the field of early childhood. Although this is done in some areas, it seems to come in and go out of favour in line with the political party in power at the time. Marmot's and Field's reports, although very informative, have come under scrutiny for driving certain political agendas at the time (Bruer 2011). However, when policy recommendations are globally agreed and actioned, there is a greater likelihood of acceptance and implementation. Unicef is successfully using neuroscience to help shape their Care for Child Development programme which supports parents to interact more and to be more responsive to their children, as well as offering the latest nutrition advice. Identified below are the recommendations by Lake and Chan (2014) which are actioned by Unicef. The influence of neuroscience is clear at a glance:

- Focusing on early interventions that start with prenatal care.
- Ensuring that policies and interventions involve health, nutrition, high-quality care-giving and protection.
- Including brain development in efforts to design effective programmes.

This will require policymakers, neuroscientists, early childhood professionals and educational researchers to work together, with close effective coordination and monitoring from each field in order to ensure that the evidence produced can be interpreted in early childhood policies and used by practitioners at an international level, consistently. Such a strategy would lend itself to raising awareness and understanding about the science underpinning the developing brain, and as a result, standardize this knowledge (Howard-Jones 2010).

According to Spelke (1999), better functional models of early childhood development are needed to help build understanding of *how* neural structures support development. When this is complemented by an increased understanding of attachment theory and its implications for child development, she believes that the use of neuromyths in early childhood policy and practice can then be minimized. In addition to this, Wolfe (2007), an educational consultant and expert on brain research, believes that the answer lies with early childhood educators to help bridge the gap between the field of neuroscience research and education. Wolfe (2007), cited in Rushton (2011: 92), explains:

> It is our responsibility as early childhood educators to understand that every child each school year represents a virtual explosion of dendritic growth. We are so fortunate to be in a profession where we can create learning opportunities to best support young children's development and their biological wiring, so let's start there.

While some readers might find this a tall order, is it such an insurmountable task? After all, it is we who have the means to take the complex terminology and abstract scientific concepts from the laboratories and give them meaning with the babies and young children entrusted in our care. We know that billions of synaptic connections are made within the first five years of life, which is partially experience-dependent. Early childhood educators are central to this exciting process, through their planning of environments and provision of experiences which encourage babies and children to explore using their senses, in their own unique ways. The messages created from the stimuli are then sent from the body to the CNS for further processing. With repetition and increasing complexity of experiences and interactions with others and their environment, learning is reinforced, with mastery of skills.

Pause for thought

1 Why do you think there is yet to be a dialogue between neuroscientists and early childhood practitioners?
2 Can you suggest two realistic ways in which this gap might be bridged by early childhood practitioners?
3 In which ways do you think that neuroscience can positively influence early childhood practice?

Top tips for practice!

- Do your homework! Take time to familiarize yourself with the brain processes involved in the learning experience of babies and young children. This can support you and your team to update your approach in light of what you find out.

- Make links with external agencies/professionals (such as those mentioned in this chapter) that do embed neuroscience in their practice. This can open up a dialogue concerning your current approach and how you might be able to make subtle changes based on the information discussed.

- Look critically at your learning environment and resources – do they encourage babies and children to engage using all their senses? Each brain responds, reasons, thinks and solves problems differently!

- Existing neural networks change in response to learning experiences – reflect on whether you (and your team) provide learning experiences that challenge the ability of the brain to respond actively, to assimilate information from a range of sources and generate new ideas.

- Do the activities and resources provide enough stimulation while encouraging problem-solving among babies, toddlers and slightly older children, equally?

Concluding thoughts

This chapter has presented some of the key issues that continue to persist in early childhood development discourse and neuroscience. Reasons as to why there is such a gap between these two fields and their impact on early childhood education and care were explored, with suggestions as to how this gap might be bridged. The references included within this chapter also highlight the complexities and possibilities of working from a neuroscientific perspective with evidence provided from a range of studies that help to avoid drawing on reductionist approaches to working with young children. These confirm that difficulties and errors will form an inevitable part of the process of incorporating neuroscience into provision of early childhood education and care, but that on the whole it is a worthwhile endeavour because neuro-science not only confirms existing knowledge concerning good practice but also extends it by providing a scientific base from which to understand the developing brain. This is recognized by the OECD (2007: 7), which confirms:

Neuroscience builds on the conclusions of existing knowledge from other sources, but the neuroscientific contribution is important even for results already known because:

It is opening up understanding of causation not just correlation …

By revealing the mechanisms through which effects are produced, it can help identify effective interventions and solutions.

The key, then, is that neuroscience *does* explain causes for certain behaviours (see Chapter 3), as opposed to making connections based on assumption only. What needs to start happening on a wider scale is the distribution of reliable and evaluated evidence that can increase practitioners' knowledge base and enhance their work with young children. As the field of early childhood education and care continues to expand, there is no reason why they should not apply the new knowledge generated by neuroscience. Hopefully you will find the questions and tips for practice useful in encouraging discussion and debate within your teams and as part of your multidisciplinary working.

Further reading

Bruer, J. (2011). *Revisiting 'The Myth of the First Three Years'*. Canterbury: University of Kent, Centre for Parenting Culture Studies.

This is Bruer's review of his book *The Myth of the First Three Years*. It is an excellent read, containing in-depth discussion concerning some of the more prevalent neuromyths that abound on the subject of early childhood development and the brain. Three of the myths explored are the concept of critical periods, enriched environments and critical period constrains regarding socio-emotional development. Some of his questions and recommendations on page 12 can be interpreted as disagreeable and consequently invite challenge, which in itself makes for healthy and informed debate among early childhood professionals and neuroscientists.

Organisation for Economic Co-operation and Development (OECD) (2007). *Understanding the Brain: The Birth of a Learning Science*. Paris: OECD.

This publication is an excellent resource which succinctly presents the challenges and possibilities of using neuroscience to inform education policy and practice. It also identifies examples of good practice where neuroscience is already being used to inform policy and practice. Although this publication does not focus on early childhood, it does cover important issues such as the importance of nurturing and the environment in early development and learning. It also discusses eight common neuromyths that are useful to familiarize yourself with in order to avoid carrying false ideas such as these into your practice.

Save the Children (2015). *Lighting up Young Brains: How Parents, Carers and Nurseries Support Children's Brain Development in the First Five Years*. London: Save the Children.

This publication is very user-friendly. It clearly outlines early brain development and its close interrelationship with language development, with practical guidance on how adults can nurture early brain and language development. Case studies and tips for good practice are included throughout, to help clarify the ideas presented.

Chapter 3
Emotional Well-being: How Can We Help to Build Healthy Brains?

Children's emotional well-being is central to their overall well-being and identity. It is part of the brain's make-up, with its development being shaped in response to the environment and the child's experiences – be these positive or negative. Sroufe (1997: 25) explains:

> Maturation of the brain, including pathways for emotion and emotional regulation, is experience dependent, that is, social interactions directly influence central nervous development.

This chapter examines the emotional well-being of young children from birth into early childhood with regard to early brain development. Current knowledge from neuroscience will be woven into the discussions to provide a different way of understanding the pivotal role of emotions in children, and how early experiences and interactions can govern emotional states and the ability to learn. Key brain regions such as the limbic system will be discussed in relation to the impact of stress on emotional well-being, followed by an introductory discussion of Polyvagal Theory and the part it can play in helping us to understand the link between the brain, heart and emotions. The importance of early care-giving and consequences of its absence will be explored, concluding with an exploration of the role of brain plasticity in overcoming early childhood adversity. *While reading, reflect on the themes in relation to your work with babies and young children, and their families.*

What is emotional development?

Healthy emotional development is essential to general well-being. Babies *expect* social interaction – becoming very upset if they do not get it (Delafield-Butt and Trevarthen 2013; Nagy and Molnar 2004). Babies are quickly able to imitate facial gestures and interpret parental responses (tone of voice, body language and facial expressions) as being affectionate or soothing or hostile and frightening. Where babies and children enjoy stable, consistently responsive and respectful interactions with primary carers, they are more likely to feel secure and happy, and consequently develop resilience and the ability to identify and regulate their own emotions as they grow (WHO 2014; DeRosnay et al. 2006).

Emotional development includes feeling and expressing emotions such as happiness, sadness, surprise, disgust and anger as well as having an understanding of one's own emotional states and that of others. Empathy, compassion and kindness tend to form a part of this repertoire for infants and adults alike. Key differences are that the emotional well-being and development of an infant requires consistent regulation, affection and support from primary caregivers in order to help them to manage and express their feelings (Thompson 2006a; Thompson and Goodvin 2005). This is all the more critical for babies as they are unable to regulate or articulate their feelings.

Pause for thought

1 When did you last take some time out to read or refer to your baby policy concerning attachment to inform your practice?
2 Reflect on your general approach to supporting infants from birth to three years during times of discomfort or distress.
   a) Have you ever found it challenging when trying to calm an infant?
   b) What strategies have you used to help you in such instances?
   c) In which ways did this enable the infant to calm down?

The statutory framework in England, the Early Years Foundation Stage (EYFS 2014) emphasizes the centrality of emotional development, placing it at the heart of the document. It is one of the three prime areas deemed:

> ... particularly crucial for igniting children's curiosity and enthusiasm for learning, and for building their capacity to learn, form relationships and thrive (2014: 7).

In this document, the influence of emotional well-being on a child's ability to learn is rightly acknowledged. When a child feels content and secure they are better able to concentrate and explore their environment, learning as they go. Conversely, when factors external to the child are not conducive to learning (such as chronic parental stress, consistently negative parental responses to the infant, depression, long-term financial difficulties, disability and substance abuse), the child might not be able to learn and thrive and will often start to externalize their feelings of distress, fear and isolation in negative ways (Twardosz and Lutzker 2010; Zeitlin 1994; West and Prinz 1987). With this in mind, emotional development does not necessarily occur in a linear fashion, nor should it be assumed that all children will easily pass through their phases of emotional development.

Consider your approach when meeting the emotional needs of babies and children: do you instinctively adapt your tone of voice, language, level of support and expectations according to individual temperaments and characteristics?

The burgeoning brain – life in the womb

The brain starts to take shape in the womb, but by no means is this process complete before birth. During the typical nine months in utero, brain growth and development is as incredible as it is rapid, with a multitude of complex factors impacting on this crucial process: maternal health and lifestyle being two critical factors. Factors such as diet, exposure to stress and drug and alcohol misuse, dependency on prescribed drugs and pollution all shape foetal development, with likely effects into adulthood (Gerhardt 2015; Barker 1995). Although greater emphasis is placed on maternal health today, with increasing evidence provided by ultrasound, fMRI and other imaging techniques, the acknowledgment of this crucial period spans centuries. The English poet Samuel Taylor Coleridge expressed the significance of prenatal development over two centuries ago (1836: 244):

> The history of man for the nine months preceding his birth would, probably, be far more interesting and contain events of greater moment than all the three score and ten years that follow it.

The foetus is not a passive occupant of the womb, but instead one which responds, grows and develops directly in response to the stimuli it receives. As a result of the brain's development in utero, the newborn baby is already capable of responding to familiar voices – 'tuning into' and preferring her mother's voice (Kisilevsky et al. 2008), distinguishing between positive and negative facial expressions and able to move in response to appealing stimuli. Research conducted by Delafield-Butt and Trevarthen (2013: 199) demonstrates this:

> A newborn infant's movements are especially sensitive to sight, hearing and touch of an attentive mother in face-to-face engagement, and they can take part in a shared narrative of expressive action.

All these interactions with the environment, and those in it, continually lead to the growth of more synaptic connections in the brain. Thus, the foundations are laid in pregnancy and are consolidated during the first two years of life. During this time of rapid growth and development, the infant's emotional repertoire and emotional 'tool-kit' are established – directly as a result of the early care-giving received.

Early on in the first trimester, the neural tube begins growing – its growth and change gradually leads to the formation of the forebrain, midbrain and hindbrain at around week six or week seven of pregnancy. The most primitive region of the brain – the brainstem – is formed first (also known as the 'reptilian brain'). Evolving millions of years ago, this is primarily concerned with survival, controlling heart rate, breathing and the fight or flight mechanism. By the end of the second trimester, the brainstem is almost entirely mature. The cortex, neurons and synapses appear, enabling the foetus to make its first voluntary movements. As a result, the brain's development is accelerated by this sensory input. Formation of the **gyri** and **sulci** (the ridges and grooves on the brain's surface) leads to the thickening of the **cerebral cortex**, and, of particular importance, myelination (the coating of neurons with a fatty sheath, which enables quick information processing) begins and continues into adolescence. Babies and children lack sufficient myelination, which makes factors such as maternal physical, mental and emotional well-being essential. Regular positive experiences and stimulation through play and multi-sensory exploration are also important in the facilitation of myelination (ZERO TO THREE 2012). As children grow, myelination contributes to higher-order brain regions that control feelings, thought and memory.

Pause for thought

1 In what ways do you think that prenatal factors such as chronic stress and alcohol and drug misuse can affect an individual in:
   a) early childhood?
   b) adulthood?
2 Describe the process of myelination.
3 Explain why myelination is important in brain development and in the nursery.

The emotional brain – the limbic system

The **limbic system** (also known as the emotional brain) is the part of the brain concerned with the *registering* and *storing* of emotional information. This includes fear, anger and happiness. Babies are born with an experience-dependent limbic system that requires substantial emotional, social and cognitive stimulation to help ensure its healthy growth (Twardosz 2012).

Below is an image of the limbic system and some of its key structures which each play a role in emotional development. *As you read, think about how each structure contributes to the formation of early childhood emotional development.*

Figure 3.1 The limbic system

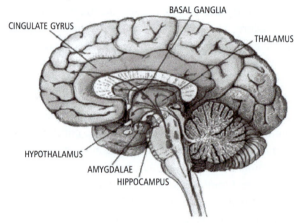

- The **thalamus** (often known as a relay station) is located at the top of the brainstem. It takes in sensory information and then passes it on to the cerebral cortex. The thalamus also regulates consciousness and sleep. In addition, it plays a role in controlling the motor systems of the brain which are responsible for voluntary bodily movement and coordination.

- The **hippocampus** is vital for memory, learning and regulating emotions.

- The **amygdalae** are two almond-shaped structures situated close to the hippocampus. The amygdalae are responsible for decoding emotions, identifying possible threats, preparing the body for emergency situations and storing fear memories. These are also responsible for the development of fear.

- The **hypothalamus** is located beneath the thalamus and is responsible for producing chemical messengers (hormones). The hypothalamus regulates physiological functions such as temperature, hunger, thirst and sleep. It also controls the release of hormones from the pituitary gland. The hormones **cortisol** and **oxytocin** are chemicals produced by the hypothalamus. (We will have more about these further on in the chapter.)

- The **basal ganglia** are the cluster of nuclei situated outside and above the limbic system. They help to regulate automatic movement, focus attention and connect the cerebral cortex with the cerebellum.

- The **cingulate gyrus** helps to regulate emotion and pain. It is also responsible for identifying fear – moving the body away from negative stimuli.

When we reflect on the limbic system, the inextricable link between the brain and emotions becomes all the more clearer – our very being is governed by our emotional states. Emotions are linked with early developing regions of the human nervous system, including structures of the limbic system and the brainstem (Barrett et al. 2007; Bell and Wolfe 2007; Thompson et al. 2003). The ability to make judgements and take risks, respond to frightening situations and learn new skills are all dictated by how secure, confident and happy a child feels. Development of the limbic system is also experience-expectant, with early experiences shaping connectivity in this brain region. This includes responsive interactions and care-giving, with regular opportunities for the child to practice using language and manage their emotions with the support of a sensitive and attuned parent or primary carer. Correspondingly, how a child manages their emotions in difficult situations shapes pathways in the brain, which in turn create the 'blueprint' for future emotional responses and behaviours (National Scientific Council on the Developing Child 2010b). Neuroimaging studies show that structures including the amygdalae, hypothalamus and prefrontal cortex are pivotal in producing emotions, memory and the regulation of behaviour and emotions (Trevarthen and Delafield-Butt 2013; Trevarthen et al. 2006). The section below explores the impact of poor care-giving and chronic adverse early experiences on cortisol levels in infants.

The limbic system predominantly works on a stimulus-response level, which means that infants can react without thinking too much first. In the nursery, this might look like a child hitting another child because they snatched a toy from their hand, or a child screaming and crying when frightened by something. It may also include a young child (particularly two-year-olds) experiencing a 'tantrum'. If you reflect on the toddlers in the nursery, most tantrums commonly occur because they get frustrated easily and have few problem-solving skills. Some have trouble asking for things and expressing their feelings. Tantrums are also most likely to occur when toddlers are exhausted, hungry or overexcited. *How* the adult intervenes can make all the difference to the child's future responses in similar situations. This is where brain regions such as the prefrontal cortex (PFC) are also important. Maturation of the PFC enables the infant to develop emotional competence as it exerts control over the limbic system and facilitates *higher order thinking* such as planning, decision-making, problem-solving and impulse control. This means that young children will be better able to think before they act. Within the daily routine this might manifest as a reduction in emotional outbursts and improved concentration during interactions and play activities.

Catastrophic cortisol – what happens to cortisol levels during prolonged stress

Stress is an inevitable part of most of our lives and can sometimes be useful in terms of motivating us into action. There are three types of stress and stress response:

- *Positive stress and positive stress response* are part of healthy emotional development, with a brief increased heart rate and a mild rise in cortisol levels. An example of a positive stress response might be an infant's response to her first day at nursery, or with a new child minder.

- *Tolerable stress and tolerable stress response* includes serious but short-term stress like divorce, bereavement or a severe injury. Where supportive relationships and networks exist, these types of stress can be overcome without causing significant long-term damage to the brain's developing architecture.

- *Toxic stress and toxic stress response* include abuse and neglect, parental substance abuse, maternal depression and the continual absence of loving, supportive relationships. These types of stress and their responses are the most harmful to an infant's emotional brain development and all-round well-being. This is due to the brain's developing architecture being disrupted, with deleterious effects on learning and development (Twardosz and Lutzker 2010; Tarullo and Gunnar 2006).

It is useful to distinguish between the three types of stress and their response in order to help you to identify an infant's type of stress response and act accordingly.

Cortisol (also known as the stress hormone) is a hormone that plays a key role in managing stressful situations. Cortisol is 'at work' in all of us. Levels are raised in the morning, which gives us our 'get up and go', eventually winding down in the evening to help us relax. Cortisol is referred to as the stress hormone because it works by shutting down essential physiological functions and increasing heart rate and blood pressure – all part of the fight or flight response – when dealing with stressful situations. So, in small doses, it is very useful in helping children and adults alike to cope with threatening or stressful situations by preparing the mind and body to fight or flee. Cortisol also plays a role in the metabolism of fats, proteins and carbohydrates and provides additional energy for muscles by increasing glucose levels in the blood.

Cortisol has a powerful effect on early childhood development. When babies and young children are continually exposed to stressful, threatening situations in the home, or whose needs for attachment and affection continually go unmet, they eventually develop a hyper-reactive stress response (National Scientific Council on the Developing Child 2010b). This means that even when there is no danger, the infant's brain and body exist in a constantly heightened state, ready to deal with the threatening situation (this is not the case for everyone, as some individuals are to a greater or lesser extent resilient to chronic stressors). This consequently places a continual strain on the developing nervous system and other vital organs which cannot function efficiently due to the constant high levels of cortisol being released. Such continuous high cortisol levels are also linked to high activity in the right frontal brain which is responsible for producing feelings of irritability, fear and responses such as withdrawal from others.

Also relevant to the discussion concerning children's stress response is the *hypothalamic–pituitary–adrenal axis* (HPA axis). The HPA axis relates to the complex interactions and effects that take place between the hypothalamus and the pituitary and adrenal glands. Its primary function is to regulate the stress response, helping the individual to respond appropriately to stressors in their environment. Figure 3.2 gives a simple overview of the interactions (or processes) involved when a child experiences a stressor. It is important to remember that acute stress experienced in the daily routine, such as going

Figure 3.2 The role of the HPA axis in stress response

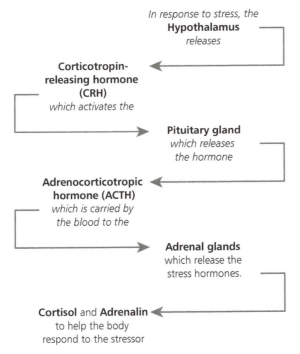

to nursery or having to practice for a special assembly, can actually prove beneficial, as it can sharpen attention and can help improve performance. The stress hormones released in stress response, such as cortisol and adrenalin, stop the body doing anything that is not considered necessary, to enable the child to put all of their resources into dealing with the stressor (or threat). However, if a child is constantly stressed or feels under threat, then their HPA axis will continually be signalling to their body to increase blood pressure and suppress important functions like immunity and digestion. Chronic or toxic stress can therefore have damaging effects on a child's developing physiological health as well as their psychosocial health.

Notably, the HPA system is not fully developed at birth and partly depends on the quality of the infant–parent attachment, early experiences and consistently supportive care-giving (Kinlein et al. 2015; Tarullo and Gunnar 2006; Fonagy et al. 2002). Infants are dependent on their parents or primary carers to help regulate their response to stress. Where this continually fails to happen, an infant's capacity to regulate themselves becomes significantly compromised. Sleeping and feeding difficulties may occur, as well as excessive crying and attachment problems in the long term due to the absence of protective and attuned care-giving (Kinlein et al. 2015; Briggs-Gowan 2006). Studies (Russo et al. 2012; Burke et al. 2005; Kaufman and Charney 2001; Larson et al. 1998) suggest that stressful early environments contribute to physiological dysregulation of an individual's stress regulation systems, particularly the HPA axis. Tarullo et al. (2006: 637) highlight the detrimental effects of chronically high cortisol basal levels:

> If basal levels of cortisol are chronically high, as has been observed in maltreated children, immune suppression, reduced synaptic plasticity and other deleterious effects can result. Moreover, chronically elevated cortisol and CRH during development, when brain circuits are still maturing, appears to shape the way these circuits interpret environmental threat and the magnitude and duration of stress responses in the future.

Ultimately, evidence provided by studies such as those cited in this section concerning the HPA axis and its detrimental effects on infant health can, and should be, used to inform neuroprotective interventions. Newman et al. (2015: 8) suggest that:

> Early identification of caregivers at risk of traumatising interactions with their children is a core strategy in the prevention of emotional and psychological difficulties in children.

Interventions can include the following:

- addressing the parent's or primary carer's own attachment difficulties.
- providing therapy for the parent or primary carer.
- reducing stress.
- encouraging reflection.
- utilizing video interaction guidance on parent–infant interactions.
- offering training to parents, primary carers or foster parents to be more sensitive and responsive caregivers.

Epigenetic changes (changes that occur to gene expression as a result of influences *external* to the infant) can take place as a result of chronic maltreatment in early childhood. These changes will be touched upon in this section and will be further examined in Chapter 5, as part of the nature versus nurture question. In essence, these changes target the neural connections that are in the process of being made, exerting a direct influence on early brain development (Newman et al. 2015). This means that neural circuits connect and embed in response to these stressors, with higher thinking functions (which are necessary for effective learning and appropriate social behaviour) significantly compromised. Newman et al. (2015: 5) explain:

> The impact of lack of early regulation and high levels of stress is potentially widespread and includes epigenetic effects, with impacts on emerging stress regulatory mechanisms as well as on attachment organisation.

It comes as little surprise that adversity early in childhood is strongly linked to all-encompassing mental, emotional and physical ill health in childhood and adulthood such as depression, anxiety and addictive disorders (De Bellis and Kuchibhatla 2006; Strathearn 2006; O'Connor et al. 2005). Sethi et al. (2013: 12) explain:

> The World Health Organisation (WHO) estimates that child maltreatment is responsible for almost a quarter of mental health disorders. Its economic and social costs are on a par with those for all non-communicable diseases, including cancer, obesity, diabetes, heart and respiratory diseases.

All this makes for a very challenging journey through life, but with the right support mechanisms put in place early enough, young children stand a fighting chance. Knowledgeable professionals and practitioners who are well versed in neuroscience and its role in educating us about the developing brain and the impact of early adversity and stress on its development can make a welcome difference in helping to change the trajectory of a child's life for the better. The following section explores brain **plasticity** and its role in enabling recovery after early childhood maltreatment.

Pause for thought

1 What are the possible effects on a child's socio-emotional well-being if they do not have responsive adults in their lives?
2 Identify two ways you could encourage your team to work in ways that are more tuned into children's emotional states (not just when they are feeling upset).
3 Explain the role of the HPA axis in a child's response to stress.
4 What strategies have you used to help reduce toxic stress in young children?
5 How far and in what ways did these strategies improve outcomes for them?
6 What are the advantages of early intervention in ameliorating the effects of early childhood adversity?

The brain and self-regulation

Self-regulation can also prove difficult for adults to master, especially when feeling low or particularly anxious and our amygdalae hastily enter into defence mode. This is particularly true for an infant with an immature central nervous system. Healthy brain development thus requires adult input that is reflective, positive and consistent. The amygdalae (singular, amygdala) play an integral part in children being able to self-regulate in times of stress or perceived danger: this is where the *amygdala hijack* can come into play (Goleman 1996; LeDoux 1991). When emotions hijack the brain, it means that the 'thinking' parts of the brain instantly become compromised and thus not able to execute their functions, as there is no time for rational thought. The thalamus and the neocortex (which receives and stores information for remembering and decision-making, and helps us to judge our responses to our surroundings) get bypassed, with a weighty and immediate reduction in memory, rational thought and planning. The result of this inability to respond rationally is that our actions are led by our emotions, not our calm, rational brain (Gunnar and Vazquez 2006; Sapolsky 1996).

Understanding which neural mechanisms are affected by the amygdala hijack and how this happens can assist us when we are trying to help children avoid the amygdala hijack from taking over during times of stress. Talking through alternative responses and modelling these with children can encourage them to use these simple strategies instead of acting on their emotions (which is easier said than done sometimes!). The pressures and stresses of modern day life is resulting in babies and young children experiencing very high stress and anxiety levels (Meadows 2016; Evans 2015). Hughes and Baylin (2012: 69) explain the possible long-term effects of childhood stress:

> Early childhood stress is now known to be a risk factor for substance abuse and addiction, and the underlying neurobiology of this risk is thought to be highly related to the early suppression of the oxytocin and dopamine systems in children.

With less time to play and many families no longer having extended networks of support to enable children to be cared for by family members, infants are instead placed in day care for up to ten hours, five days per week. Little time with key family members means that very young children often have to find their own ways of managing stress. In a hectic early years setting this can further add to a young child's distress if practitioners misinterpret signs of stress as 'poor' behaviour. These signs will differ according to the age of the child, which means practitioners need to be knowledgeable about stress in young children and how to manage it. This knowledge can inform practitioners' interventions as well as help them minimize the presence of stressful situations for babies and young children. Effectively managed stressful moments can thus be turned into opportunities for young children to establish self-calming skills. Young children respond very differently to stress – what is perceived as stressful to one child, may not seem so bad to another. What is vital is practitioners' ability to show empathy, patience and respect when a young child is experiencing overwhelming feelings in response to the trigger. When an adult fails to empathize with the child and instead resorts to shouting or quickly issuing punishments (commonly 'time out' or threatening to withhold treats), this serves to further anger the child and does nothing to tackle the cause of the problem – what has led the amygdala to go into overdrive.

Proceed with caution!

As with all theories stemming from neuroscience, the amygdala hijack also needs to be considered carefully. While the idea of our amygdala taking over during times of perceived threat and stress is accepted, the idea that this is all the amygdala does has led to the neuroscientist behind the amygdala hijack responding to its rarely questioned, celebrated status within education, psychology and business circles. In his blog, *I Got a Mind to Tell You* (15 March 2016), neuroscientist Joseph LeDoux explains at length how findings from neuroscience can easily be misinterpreted – especially in the case of the amygdala hijack, which got lost in translation. While the amygdala is involved in the fear response, it is not the sole brain region involved in fear processing, yet its subsequent representation (Goleman 1996) has resulted in mass recognition, particularly in the US and the UK. LeDoux (2015) clarifies the situation:

> The idea that the amygdala is the home of fear in the brain is just that
> – an idea. It is not a scientific finding but instead a conclusion based on
> an interpretation of a finding. Conscious fear is a product of cognitive
> systems in the neocortex that operate in parallel with the amygdala circuit.
> But that subtlety (the distinction between conscious and non-conscious
> aspects of fear) was lost on most people. This problem is especially acute in
> neuroscience, where we start from mental state words (like fear) and treat
> the words as if they are entities that live in brain areas (like the amygdala).

So perhaps the issue is more about the captivating but imprecise labels that are given to describe the neural mechanisms at work when children feel threatened or anxious. After all, it is easier to remember the term amygdala hijack than the lengthier explanation which actually breaks down what happens to all the concerned brain regions when children are scared.

The concept of the amygdala hijack is used commonly by professional coaches and leadership trainers (to name but a few) in their influential presentations which give their position validity and immediate kudos due to the science informing their explanations. Although this is not ideal, it can be viewed positively to an extent because theories like the amygdala hijack gain mass attention, being adapted for education. For example, early childhood practitioners, trainers, lecturers and management do occasionally refer to the amygdala hijack when discussing the effects of stress, anxiety and maltreatment, and their long-term consequences for very young children. This does not necessarily mean that we are being misled, but we may be missing out on the whole picture. Below is a case study provided by a nursery school teacher in North London.

Case study

In a busy classroom of thirty children aged between three and five years, we do a lot of talking about feelings in small groups and on a one-to-one basis. Children talk about ways they could manage their emotions *before* they get angry. Some of their ideas have included walking away from the situation, kicking a ball in the playground and speaking with a teacher. Obviously this does not always happen! But it's worthwhile talking through their ideas so that they know there are alternatives to 'losing it'. This regular time with the children helps us to identify the ways in which they are likely to respond when they get angry but also helps us to prevent them from getting angry in the first place. It's about identifying their triggers. We *connect* with every child from the time they start and really get to know them; their personality, whether they're easily upset or resilient to the fast tempo of school life. When children do experience stress and anger, we always stay calm when we intervene. This is easier for some staff than others, but staying calm and speaking to the child very calmly and softly helps them to tune into what you're saying and reassures them that there is no real danger. Our aim is to always calm the child down as quickly as possible to minimize their distress, while acknowledging their feelings and giving them time and space to express their feelings in a safe way. We never threaten with consequences as this is wholly destructive.

As a team, we regularly share articles on the effects of stress on the brain and talk through the advice given which we adapt and try out. Our aim is to eventually enable the children to overcome feelings of anger independently, in ways that do not harm themselves or others.

Pause for thought

1   Explain the role of the limbic system in an infant's emotional development.
2   Discuss one way in which practitioners can help to promote healthy development of the limbic system.
3   Which situations do you see commonly trigger the amygdala hijack in young children and staff?
4a Which strategies do you find useful when supporting a child experiencing the amygdala hijack?
4b Explain the reason behind your choices.
5   How does understanding the brain help us manage stress in young children?
6   Have you reflected any of this in your personal, social and emotional development (PSED) policy?

Nurturing relationships – how neuroscience is used to inform attachment aware practice

Perhaps the most powerful factor when it comes to raising children is the role of parents and primary carers. Where this role does not meet a child's needs for security, affection and consistency of affirmative parental responses, practitioners can make a positive difference to young children's self-esteem and relationships. Ameliorating the effects of poor nurturing in early childhood is just one example of how practitioners facilitate the emotional development of young children who have suffered from early childhood maltreatment and insecure attachments. Perhaps the forerunner of such approaches is the Marjorie Boxall nurture groups (1969). Boxall, an educational psychologist, designed the nurture groups in response to the number of children who, as a result of early childhood neglect and poor attachment to parents, were unable to thrive emotionally, socially and cognitively. The groups were designed using attachment theory as the framework for understanding and responding to children's behaviour. Boxall (2002: 2) explains that the nurture groups aim to:

> [c]reate the world of earliest childhood; building the basic and essential learning experiences normally gained in the first three years of life and enable children to fully meet their potential in mainstream schools. The emphasis within a nurture group is on emotional growth, focusing on broad-based experiences in an environment that promotes security, routines, clear boundaries and carefully planned, repetitive learning opportunities.

Note here the emphasis on the period between birth and three years. This does vary depending on the publication, as some researchers advocate a period up to five years. That said, we have become well versed in the discourse concerning the unique phase of *early childhood* concerning learning and development, due to its promotion in various health campaigns and child development publications. It is during this time that the brain is most malleable and thus capable of growing neuronal pathways in response to experiences. The link between this significant period and attachment aware practice is therefore encouraging, as early investment in the early years can have lifelong benefits for personal, social and emotional well-being. Nurture groups are not mandatory practice, but research shows that early childhood settings and schools across the United Kingdom are increasingly using this practice successfully (Ofsted 2011; Cooper and Tiknaz 2007). These groups are helping to bring about positive change in children who display social, emotional and behavioural difficulties, with results in improved relationships at home as well as improved educational outcomes. In their updated and revised guise, nurture groups are now informed by neuroscience and knowledge concerning early brain development. Research taken from fMRI studies is gradually beginning to inform educators about the brain's responses to poor attachments, persistent, toxic stress and threatening situations. We know all too well that when a child does not have a safe emotional base from which to understand and experience the world, their social, emotional and cognitive development becomes adversely affected. Immordino-Yang and Damasio (2007: 1) explain:

> Recent advances in neuroscience are highlighting connections between emotion, social functioning, and decision making that have the potential to revolutionize our understanding of the role of affect in education. The neurobiological evidence suggests that the aspects of cognition that we recruit most heavily in schools, namely learning, attention, memory, decision making, and social functioning, are both profoundly affected by and subsumed within the processes of emotion.

> ## Pause for thought
>
> 1 What are the links between attachment and educational attainment?
> 2 How far do you think nurture groups informed by neuroscience can improve behaviour?
> 3 What attachment aware practices are already in place in your setting?
> 4 Does your setting use findings from educational neuroscience to inform practice concerning the emotional and cognitive development of children?

Top tips for cultivating emotional competency in challenging situations

- Reflect on *your* responses and general behaviour when things do not go according to plan. Everything you do is **internalized** by children, so think about how you can adjust your responses if children are being destructive (such as swearing, shouting and displaying other forms of anger).
- Avoid punishing the child if and when a tantrum occurs. Remember that this is the amygdala hijack in action!
- With encouragement from their carers, young children can learn to ask for help, go somewhere to cool down or to try different ways of doing something.
- Time and energy is well spent teaching young children *how* to manage their responses to stressful situations, instead of punishing them for not having better developed skills.
- Make the effort to role model qualities such as patience and self-control so that the young child can learn to adopt such qualities when dealing with challenging situations.
- Know the children's stressors and try to prevent these from occurring – don't forget to share successful strategies with primary carers.

Emotional development – setting positive foundations

Learning to manage one's emotions occurs over time as a result of complex interactions both within and external to the infant. What is of great importance during this time of rapid growth and change is *how* the infant's varied (and often intense) emotional states are recognized and nurtured by significant adults. These responses help create the neurobiological foundations for emotional development and the way the infant learns to manage their own emotions. Furthermore, each infant's neurobiological markers are unique, based on their individual experiences and temperament. This is identified in the National Scientific Council on the Developing Child (2011: 2):

> The emotional health of young children is closely tied to the social and emotional characteristics of the environments in which they live.

Neurological changes in the infant occur in response to care-giving interactions with the primary carer such as playing, feeding, comforting, holding and communicating (Schore 2001; Cicchetti and Tucker 1994; Spangler et al. 1994). It barely needs restating, then, that caregivers play a vital role in the infant's neurobiological regulation of emotions. Where caregivers soothe the infant when they show distress or cry, this helps infants to self-regulate uncomfortable and negative emotions and consequently reduce the level of stress hormones such as cortisol (Gunnar and Quevado 2007; Gunnar and Davis 2003).

Polyvagal Theory (PVT) and the polyvagal system (PVS) – why it is useful to understand these as part of the early childhood discourse

Polyvagal Theory was developed by Dr Stephen Porges (1995). Polyvagal literally means many (poly) nerves (vagus). PVT emphasizes the link between the brain, heart and emotions. It is concerned with the brain and the nerves involved in processing emotions. Porges (2007: 118) explains:

> The Polyvagal Theory encourages a level of inquiry that challenges scientists to incorporate an integrative understanding of the role neural mechanisms play in regulating biobehavioural processes.

Image 3.3, below, shows the face–heart connection via the polyvagal system.
    The nerves that control the facial muscles and vocalization are linked in the brainstem by the myelinated vagus nerve.

Figure 3.3 The polyvagal system

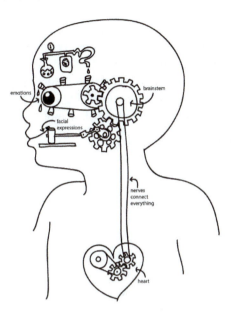

Within the PVS, the vagus nerve connects the brain, facial muscles, heart and gut, and manages the three general states in which humans exist, these being:

1 Social engagement
2 Fight or flight
3 Frozen

So when, for example, a child's nervous system feels threatened or detects risk and fear, and is unable to use its highest level strategy – *social engagement* or *mobilization strategies* – it cannot function normally. This results in physiological changes such as increased heart rate and blood pressure, rapid breathing, increased sweat production and a slower working digestive system as blood supply is diverted to more critical areas. Porges (2007: 119) observes:

> The myelinated vagus actively inhibits the sympathetic nervous system's influences on the heart and dampens hypothalamic-pituitary adrenal (HPA) axis activity.

The polyvagal system consists of two main systems in the body, these being the autonomic nervous system (ANS) and the central nervous system (CNS). Porges' insights enable us to understand the function of the nervous system when trying to regulate itself and protect us from danger. When a child feels safe and secure, their social engagement system (face, eyes, mouth and middle ear) and heart work 'happily' together and the child can interact with others, engage in learning and is generally able to function 'normally'. Unfortunately some children are programmed from an early age to operate in fight or flight mode. Reasons such as early childhood trauma, abuse and chronic stress can impede children's social engagement system, leaving them in a state of continued high alert (Schaaf et al. 2010). These two opposing states are depicted below in Figures 3.4 and 3.5.

Figure 3.4: Healthy functioning polyvagal system

Figure 3.5: Unhealthy functioning polyvagal system

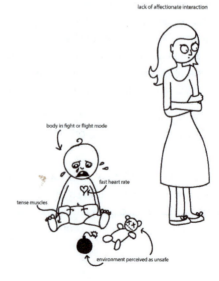

When children are not able to work from their social engagement strategy, they instead revert to a modified fight/flight strategy, which puts them in high alert. If children use too much of the fight/flight or freeze strategies, health issues may result, as bodily functions become compromised due to functioning in fight/flight mode. This all-important neural connection between the brain and heart was recognized by Charles Darwin (1872: 1), whose observations long preceded those of Porges:

> When the heart is affected it reacts on the brain; and the state of the
> brain again reacts through the vagus nerve on the heart; so that under any
> excitement there will be much mutual action and reaction between these,
> the two most important organs of the body.

According to Porges, the polyvagal system is directly impacted by social and emotional interactions from birth. Consider the close relationship between a parent (or primary carer) and their baby. In theory, during such intersubjective interactions their deep connection and processing of each other's emotions is conveyed through facial expression, eye contact, tone of voice and hand movements. When these interactions are affirming and affectionate, they exert a soothing and calming effect on the baby by causing an involuntary change via the ANS – that is, heart rate is slow, the muscles are relaxed and the body can function normally as the baby feels safe.

Pause for thought

1 In your own words, describe the polyvagal system.
2 Explain how the polyvagal system influences an infant's processing of emotions.
3 What is the role of touch in soothing an infant?
4 In what ways might Polyvagal Theory be useful to early childhood practitioners and primary carers?

The power of love – the importance of positive early care-giving

Life for a baby is understandably very demanding and unpredictable. Entering the world completely unable to care for themselves, they are dependent on primary carers to meet their every need to ensure survival. Adjusting to life outside of the womb can hence be very overwhelming. When primary carers respond to their baby's cries of hunger, tiredness or discomfort and consistently show love, the infant not only has his basic physiological needs met but these experiences are being laid down as neural pathways. Gerhardt (2015: 82) informs us:

> The kinds of emotional experience that the baby has with his caregivers are biologically embedded. They get written into the child's physiology because this is the period of human life when regulatory habits are being formed.

It stands to reason that emotions play a crucial functional role. They regulate psychological states and they control interpersonal and social interaction. A child's ability to demonstrate appropriate interpersonal skills and manage social interactions greatly depends on the quality of their early experiences within their primary socialization. For example, where a child is 'coached' by their primary carers and staff at nursery to recognize and talk about their feelings from an early age, they are more likely to grow up emotionally literate. Emotions such as fear, self-doubt, sadness and frustration can be talked about without the child feeling stigmatized. Generally, when babies and very young children understand that they are being listened to (as opposed to being told off for trying to express their feelings), the intensity of their feelings begin to dissipate. This results in calmer babies and can create an opening for joint problem solving between slightly older children and their primary carers.

Pause for thought

1a What advice would you give to a parent who is finding it difficult to manage her one-year-old daughter's displays of anger?

1b How would you convey to this parent, the importance of managing these moments positively, in terms of nurturing her daughter's developing emotional skills?

2 Reflect on your setting's strategies for encouraging emotionally literate behaviour. How could these be improved in light of current information concerning early brain development?

3 Do you think that practitioners are sufficiently educated in strategies to help parents to avoid or ameliorate 'toxic' stress?

What happens when early care-giving does not take place?

Most parents experience 'off-days' – when their emotional resources are low and resultant responses towards their children are a little below par. This is understandable given the many and complex challenges that life has in store. However, if the baby or young child only ever experiences negative feedback from his primary carers it becomes a serious issue because the young brain develops in response to the care received. Connections are made and strengthened as a result of early experiences, which is why there is so much emphasis on the *1001 days* (the period from conception to the child's second birthday). The 'serve and return' dynamic (where a baby's cries, babbles and expression of emotion are responded to affectionately) therefore plays an integral role in supporting healthy emotional, social and brain development.

The quality of attachment is negatively affected where infants endure a relationship with their primary carers that lacks in warmth, affection and responsive care. For example, some families still adhere to the old, cruel adage 'spare the rod, spoil the child' and implement discipline strategies that do more harm than good. Hitting a child for toilet accidents, for showing humour or being 'cheeky' or not doing as told by parents still occurs. Such beliefs and harmful approaches to child-rearing are difficult to 'undo' as they are often entrenched culturally and passed down over generations. Corporal punishment is an outmoded, mean and ineffective method of discipline and should have no place in any society.

Babies and young children learn from their parents, and where a culture of hitting, belting or smacking is the norm during times of parental stress, they too will learn that this is the way to deal with unwanted behaviour (Hughes and Baylin 2012). Other detrimental effects of harsh or physical punishment include insecure attachments, inability to concentrate or to experience positive emotions, immune disorders and depression. This can debilitate young children and continue well into adulthood (Depue et al. 1994; Davidson and Fox 1992). Gerhardt (2015: 149) informs us of the impact of neglect and physical abuse on the developing brain:

> In depression there is a reduced density of neurons in the dorsolateral
> part of the prefrontal cortex, the area that develops in toddlerhood and is
> involved in verbalising feelings. The more depressed you are, the less activity
> there is in the prefrontal cortex.

Gerhardt here emphasizes how the prefrontal cortex is significantly compromised if a child isn't happy, or continually encounters prolonged stress. It is a key part of the brain concerning emotion regulation and reflective functioning (Hughes and Baylin 2012). Hughes and Baylin (2012: 145) explain the benefits of temporarily putting personal needs aside in order to be able to understand the infant's cues and respond sensitively to their needs:

> By using your executive abilities, you are able to reduce or manage your emotional intensity sufficiently to focus on your child's needs and, if necessary, put aside what you feel like doing. Emotional regulation isn't an end to itself; it's a step in a multistep dance of staying parental when we are experiencing inevitable conflicts between our parental and our unparental reactions and motivations.

'Is all lost?' – the role of brain plasticity in overcoming early childhood adversity

We now know that neurons that wire together fire together – every time we interact, reflect, take exercise, eat, sleep and experience a range of emotions, neurons become activated and strengthened as they make an increasing number of connections with each other. The brain's capability to grow and change in response to stimuli does not end in childhood; instead it continues throughout adulthood. Known as plasticity, this is at last part of the early childhood discourse with regard to evidencing the impact of early maltreatment in shaping brain development. By contrast, evidence from neuroscience spanning over two decades also confirms that the brain continues to make new connections throughout life, but that the extent of a brain's plasticity is dependent on the stage of development and the particular brain system or region affected by the maltreatment (Perry 2006).

One example of early childhood chronic maltreatment is the tragic case of the Romanian orphans who were left in utter sensory deprivation in orphanages under the Ceausescu regime during the late 1980s and early 1990s. These infants existed in buildings that had little lighting and heating if any; infants were strapped to cots, deprived of emotional support, affection, stimulation and adequate nutrition. The lack of affectionate care resulted in them using repetitive self-soothing behaviour such as rocking and banging their heads against walls. Those who showed signs of distress were subjected to further abuse by staff.

In a retrospective study, Chugani et al. (2001) demonstrated comprehensive neurological abnormalities in the Romanian orphans approximately ten years post-adoption. PET scans showed decreased metabolic activity in the orbital frontal gyrus, the infralimbic prefrontal cortex, the amygdalae, the head of the hippocampus, the lateral temporal cortex and in the brainstem (Perry 2008). Developmental disorders were also commonly seen among the orphans such as autism, hyperactivity, impulsivity, language and fine and large motor delays, dysphoria (a general state of unease and deep dissatis-faction) and disorganized attachment (Perry 2006).

Recent neuroimaging studies show that those who were adopted by Western families before six months old made better progress developmentally than those who were adopted between six months and two years old (Read et al. 2001). However, studies concerning all of the adopted orphans at approxi-mately nine years upwards consistently show that key areas concerning emotion regulation such as the amygdalae, hippocampus and left orbitofrontal cortex remain underactive, with abnormal pathways between them (UNESCO 2007; Eluvathingal et al. 2006). Neuroscience enables us to identify specific brain regions that become damaged through childhood maltreatment, and how this damage affects a child cognitively, socially and emotionally. The research also consistently demonstrates the value of early intervention in supporting the brain's recovery following the maltreatment. This valuable information should be consistently used nationally to inform targeted early intervention as well as the early childhood discourse when it comes to planning curricula frameworks and health care for expectant mothers. Continual professional development programmes that inform early childhood practitioners and related health professionals can also contribute to building and consolidating understanding of neuroscience and its applications.

Pause for thought

1 What is the role of brain plasticity in overcoming early childhood maltreatment?
2 Based on your experience, do you think that teachers, care workers and other early childhood practitioners are aware of the factors that put young children's brain development at risk?
3 In your opinion, do early childhood education and care policies acknowledge that there are sensitive periods in brain development, including the role of support for mothers and their babies during these sensitive periods?

Case study

Below is an example of how one trauma parenting specialist and trainer draws upon neuroscience, childhood and adolescent brain development and attachment to inform her work with families and children affected by abuse and trauma. The approach she describes has not yet been formally evaluated in a research study nor has an objective evaluation been carried out. This case study does show the benefits she and those she works with report for taking this integrative approach, but it is important to note that what works in this instance might not work for others.

For over two decades I have worked directly with children, young people, parents and carers and whole families with complex needs relating to experiences of early childhood trauma. I have done this as a domestic violence parenting worker, a social work assistant, a children's services practitioner in a family centre for the NSPCC, as a childminder, foster carer, and as a home to school support worker. I don't have a formal qualification in trauma and am not a psychologist, neuroscientist, social worker or psychotherapist. What I do have is direct experience of doing the work and eleven years of self-study of trauma, brain development and attachment. I use my direct and learned knowledge to support parents on a 1:1 basis to have an understanding of their own and their children's behaviour and needs in the context of some level of early childhood trauma. I also train professionals in understanding the impact of early childhood trauma in a range of contexts, early years, school, supporting child to parent abuse, and parenting. I also regularly speak at conferences, again using this foundation of childhood trauma in a broad range of contexts as it underpins all that is troubling our children and society.

As soon as I began reading about how the brain is shaped by trauma I was hooked! It has helped me make sense of all that did not make sense! Understanding that our body and brain are shaped in our earliest moments and what this means for our mental and physical well-being and our ability to access our full potential, can make a real difference to how we relate to young children. As a practitioner in pre-schools and family centres I often puzzled about why children and adults often repeatedly did things which made life worse for themselves in the short and long term. I sometimes struggled to understand why some adults were so set on 'self-destruct' or a course which would mean their children would be removed from them, sometimes permanently.

Once I began reading about how the brain is shaped by early trauma and what a legacy this is for an individual, everything changed both in my practice and approach and most importantly in my thinking. I strongly believe that so much can be done to transform a child's onward journey if practitioners have their trauma and attachment glasses on when looking at a child's behaviour and development. In everything I do I offer a simplistic explanation of brain development and how this can affect daily living and relationships. I offer this to children, young people, parents, carers, professionals and anyone who will listen as I believe a basic understanding demystifies so much of human behaviour and shows it for what it truly is – an attempt to survive. Families often respond really well to understanding that there is not something 'wrong' with them as individuals but that it relates to something they could not control at the time but which can be addressed now. Having a different way of thinking about difficult behaviour – so for an adult that may be 'losing it' with a child, or for a child that might be lashing out, for a young person it may be threatening a parent or not following instructions or requests – can be really helpful.

Once the problem becomes less about the child being the problem but more about understanding why, then I find adults are more able to view them differently. However, some adults are so wedded to the belief that children are manipulative, will be naughty unless kept on the straight and narrow through a reward and consequence based approach, it does not convince everyone. Some people are scared to learn that early exposure to overwhelming levels of fear and stress has shaped their brains and/or that of the child they are raising or caring for, so it has to be done at the level and pace of the individual. In fact I find children and young people get the idea that they need to focus on calming their survival system down in order to get on with the lives they want much more readily in most cases.

Ultimately it offers families a different way of relating to and with each other. Understanding that their brother, child or parent is in 'survival mode' so needs help, space and/or time to feel calmer so they can think and behave differently can make a real difference. It sounds easy but it can be very challenging for a traumatized brain to get and stay calm. So, working with families solely on this can take time, as it is about reprogramming their beliefs and their brains.

Top tips for practice!

- When planning provision in your setting, bear in mind that different parts of the brain may be ready to learn at different times.
- Within your teams (or individually), reflect on the key elements that are necessary in providing a nurturing emotional and social environment.
- Reflect on your provision for 'quiet time' during the routine. Is sufficient time and space created in the setting to encourage young children to pause and reflect, away from distractions?
- Consider your setting's training and support programme for staff. Are there regular opportunities to access support concerning their ability to be reflective about children rather than reactive to their behaviour?

Further reading

Gerhardt, S. (2015). *Why Love Matters*. London and New York: Routledge.

In this second edition, psychoanalytic psychotherapist Sue Gerhardt provides a compelling account of why early attachments and positive early experiences are crucial in the shaping of all-round development. She supports her argument with references to current findings from neuroscience, which prove how these early experiences set down the neurobiological patterns in the brain. A 'must-read' for parents, carers and early childhood professionals.

Hughes, D. A. and Baylin, J. (2012). *Brain-based Parenting: The Neuroscience of Caregiving for Healthy Attachment*. London and New York: Norton and Co.

Written by an attachment specialist and clinical psychologist, this book clearly and empathically details the neuroscience behind healthy care-giving and the highly complex connection between the brain and body. Real-life case studies are included to demonstrate how parents' responses to their children have as impact on neural mechanisms and how these manifest as behaviour. The concept of blocked care is also explored, with practical advice on how primary carers can work through the problem. This book is a highly worthwhile resource for primary carers, teachers, early childhood practitioners and students.

Lewis, T. M. D., Amini, F. M. D. and Lannon, R. M. D. (2001). *A General Theory of Love*. New York: Random House.

Written by three psychologists, this philosophical book explores the neuroscience of attachment. The authors examine the close relationship between our emotional well-being and the effect this exerts on our daily lives and our relationships with friends, family and children. They provide an engaging account of their years of research into neuroscience in the hope of discovering what it could reveal about relatedness, attachment and love, which they describe as gathering information derived from the 'cross-pollination of a panoply of disciplines (2001: 13), including neonatology, brain development and experimental psychology. The questions they ask give plenty of food for thought for anyone interested in what motivates their attachments.

Chapter 4
Children's Language and Communication Development: What Can Neuroscience Tell Us?

Chapter outline

What this chapter is about (p. 68)

Why you should read this chapter (p. 69)

How neuroscientific investigations can add information that will help in practical terms (p. 70)

The influence of the home learning environment (p. 83)

What this chapter is about

This chapter examines topical issues that commonly have a significant impact on young children's language and communication development. Following discussions with early childhood practitioners and teachers in a range of early childhood settings, recurring themes began to emerge which provide the focus for this chapter. These being:

1 The influence of the home learning environment
2 The role of intersubjectivity in nurturing early language and communication development
3 Autism

These three themes will be explored alongside related issues, including the role of neuroscience in helping us to understand language difficulties in young children, with clearly explained examples of technology used in these processes. The chapter will also take an objective look at the some of the evidence provided by neuroscience, as well as the challenges of using brain imaging technology to enhance understanding of the young brain. This section is useful in encouraging you to consider the advantages and the limitations of the role of neuroscience in understanding the neural mechanisms that cause language development.

The possible influences of neurological impairment on language and communication will be examined with accompanying evidence from neuro-scientific studies. This will be followed by a look at some of the current thinking in language and communication development and how evidence from neuroscience research studies can help to enhance understanding and inform the support provided for children who have language and communication difficulties.

This chapter contains useful references throughout, but in some cases these are not discussed in detail, so as to keep to the focus of the book. The references can be followed up by referring to the Bibliography. Questions are included to encourage reflection on practice and to strengthen understanding of the issues presented, with a summary of points for practice towards the end of the chapter. A list of further reading ends the chapter, to support further investigation of areas of personal or professional interest.

Why you should read this chapter

This chapter is designed to inform early childhood practitioners, lecturers and students about the nature of some commonly encountered issues concerning language and communication development, alongside compelling evidence from neuroscience. The evidence cited will be explored in terms of its practical use when supporting the development of language and communication of babies and young children. The ability to communicate effectively is a lifelong skill, with its foundations laid in early infancy. Mastering the building blocks of language is therefore a critical aspect of child development (Kuhl 2009; LeDoux 2003). Steven Pinker (1995: 12) offers an interesting explanation of how language develops in a child, likening the process as instinctive:

> Language is a distinct piece of the biological makeup of our brains. Language is a complex, specialized skill, which develops in the child spontaneously, without conscious effort or formal instruction.

The tips for practice in this chapter can be discussed within your teams and actioned as you deem appropriate, while remembering the importance of adapting them to meet children's individual needs. *The topics discussed throughout this chapter are not to be taken as 'the whole story'. They serve as starting points to help guide your understanding of some of the pertinent issues surrounding early language development and neuroscience.*

Early childhood practitioners are at the forefront of supporting the development of all babies and children entrusted in their care – particularly those whose development does not follow the 'typical' pattern expected for their age. Where there is a concern, practitioners would contact other professionals who observe, give advice and draw up ideas for interventions. This makes the issue of knowledge, competence and confidence critical. Parents often seek advice from their child's key person or other staff in the setting, in order to gain more understanding about their child's condition.

There exists a wide range of issues that can affect speech, language and communication in the short or long term. These include otitis media ('glue ear'), stammering, autism and attention deficit hyperactivity disorder (ADHD). When considering whether a young child has language difficulties, early childhood practitioners might instinctively feel that something is not quite right. This is where a deep knowledge of child development comes into effect: for example, we know that children aged three years generally love to socialize, sing, make up words, engage in conversations and use their imagination to make up stories. Some children, however, are not able to fully enjoy the experience of communicating with other children and adults around them. Being able to identify any deviations from expected development and knowing what types of behaviour could be cause for concern can have a positive impact on the education and care being offered.

Neuroscience is continually improving with brain imaging study techniques being suitable for use with infants from birth. This means that research evidence concerning how children learn to talk – and of great significance, the cognitive and social processes involved – can provide valuable information on the importance of early interactions in facilitating early language. Kuhl (2010: 1) observes:

> Neuroscience on early language learning is beginning to reveal the multiple brain systems that underlie the human language faculty.

The 'multiple brain systems' refers to the different structures of the brain (**cerebrum**, **cerebellum**, **brainstem**, **pituitary gland** and the **hypothalamus**) and the role each part plays in early learning and language development. Now that neuroscience is enabling a closer look at these systems via brain imaging studies, we can identify exactly what can go wrong and where in the brain, and the effects that atypical brain development has on language and communication.

How neuroscientific investigations can add information that will help in practical terms

The continual evolution of brain imaging studies and other related non-invasive techniques cannot be underestimated. It is due to such techniques that exciting new findings are being made that can impact on early childhood education and care. That said, unless these findings can be utilized both theoretically and practically by those working with children, they will prove limited in their applicability. *So, what are some of these findings and how might they be applied to practice?*

The role of neuroscience in helping us to understand language difficulties in young children: Where are we?

This is a very exciting time in the field of neuroscience (Jiang et al. 2012; Kuhl 2010; Gopnik 2009), enabling us to not only understand *what* babies and young children are capable of in terms of their ability to communicate, but *how*. The period from birth to three is one of remarkably rapid change (Meadows 2016; Meltzoff et al. 2009; Dowling 2004), which means that babies and young children need the input from sensitive adults to guide them through making meaning of words and, later, in the art of conversation. As Dowling (2004: 65) says:

> It is tempting to suggest that neurons can gain or lose territory, synapses rearrange and new ones form, depending on language experience. By the age of six months to a year, neural circuits have formed to discriminate and make all language sounds and to acquire grammar. If the circuitry is not used, it is rearranged or lost.

Those of us who work with children enjoy observing them play and interact as they navigate their way through relationships, but few of us outside of the science laboratory know exactly what happens in the brain as this takes place. Dowling (2004: 4) expresses in no uncertain terms the pressing need to utilize findings from neuroscience to better inform parenting, education and care:

> The challenge of understanding how the brain develops and how that understanding might help in raising the next generations to the best of our and their abilities is key to the future of humankind.

The ability to communicate effectively is a lifelong skill, with its foundations laid in early infancy. The ability to master the building blocks of language is therefore a critical aspect of child development (Trevarthen and Delafield-Butt 2013; Kuhl 2011; Pinker 2007). Today, neuroscience is providing mounting evidence for this, with insights made into developmental disorders due to a range of non-invasive techniques (Kuhl 2010), some of which are identified below.

- **Event-related Potentials (ERPs)** are commonly used in the study of speech and language processing in infants and young children.
- **Magnetic resonance imaging (MRI)** measures in young infants identify the size of various brain structures. Evidence shows that these measures are linked to language abilities later in childhood.
- **Magnetoencephalography (MEG)** shows phonetic discrimination in newborns and infants in the first year of life.

Further examples of extensively used techniques that enable us to understand how babies' and young children's brains are wired to develop language are discussed below. These examples range from safe non-invasive experiments in utero to babies and infants in social contexts. *As you read, reflect on how this information can help to build your understanding of the optimum factors that shape early language development.*

Face-to-face neuroscience (also known as functional near-infrared spectroscopy, fNIRS)

Functional near-infrared spectroscopy identifies the neural processes that are involved in face-to-face interactions between infants and adults by defining activity in specific brain regions. This is achieved by constant monitoring of haemoglobin levels in the blood (Lloyd-Fox et al. 2009; Aslin and Mehler 2005).

Advances in technology combined with better understanding of brain function have meant that the cheap and effective fNIRS system is now successfully used to track the brain activity of babies to examine how they process language (Bortfeld et al. 2007; Taga and Asakawa 2007). In practical terms, this is very exciting – as the studies not only confirm that interaction exerts a positive impact on the child's well-being, but that babies clearly demonstrate a preference for prosodic sounds when listening to speech (higher pitch and slower articulation of words than is normally adopted by adults). In line with this, MEG and EEG studies have shown a positive relationship between listening to motherese and increased attention and speech perception in infants (Bazhenova et al. 2007; Berger et al. 2006; Orekhova et al. 1999; Stroganova et al. 1998).

Used as part of trans-disciplinary working, these findings can help to improve public awareness about the special role of motherese in supporting infant communication and language development. However, children with autism demonstrate weaker connectivity between the left and right hemispheres of the brain, responding differently to language (Zhu et al. 2014). This may result in some of those symptoms attributable to autism such as communication difficulties, a repetitive pattern of interests and behaviours, language impairment and compromised social skills. This is discussed in greater detail in the section on autism.

Using cognitive neuroscience to understand children's language development

Cognitive neuroscience is integral to developing understanding children's (typical and atypical) language development. As neuroimaging techniques continue to advance, researchers are able to provide detailed illustrations of the structural changes that take place in the brain during development. Combined with disciplines including philosophy, psychology, sociology, genetics, affective and developmental neuroscience, the biological and environmental influences on brain and mind development are also examined within cognitive neuroscience.

One fascinating current study (Huth et al. 2016) using fMRI has shown how the human organizes language. The brain blood flow of six adult research participants was measured as they listened to people recount autobiographical experiences on a radio show. From the evidence generated, scientists created a comprehensive brain map showing which regions of the cerebral cortex respond to different words. (Words were grouped under headings such as emotional, social, violent, numeric and visual.) The study proved to be an effective method for mapping functional representations in the brain, which could have practical implications for supporting adults and children with communication difficulties. One exciting possibility includes clinicians tracking the brain activity of patients who have difficulty communicating and matching that data to semantic language maps to determine what patients are trying to express. The researchers acknowledge that although successful, the study needs to be replicated using a larger and even more diverse sample in order to draw more detailed conclusions. This can hopefully lead to wider dissemination and utilization of findings, which can be used to inform the work of those who work with children who have language and communication difficulties. As Huth (2016) explains:

> This discovery paves the way for brain–machine interfaces that can interpret the meaning of what people want to express. Imagine a brain–machine interface that doesn't just figure out what sounds you want to make, but what you want to say.

Using brain rhythms to identify intellectual abilities during early language acquisition in infants

Brain rhythms are also known as brain oscillations or the more familiar term 'brain waves'. These are measured using MEG and EEG to ascertain the rate of brain activity when infants are exposed to language and how much mental effort (for example, concentration and memory) they use when processing new words (Bosseler et al. 2010). Techniques like MEG and EEG can provide clearer understanding of how babies and young children process information, particularly during language acquisition, and what causes some children's brains to expend more energy (that is, to work harder) when processing the same information. Kuhl (2010: 715) draws our attention to the benefits of using brain rhythms, as opposed to behavioural techniques in examining intellectual abilities, particularly during early language development. Kuhl uses the term 'neural efficiency' when describing brain activity. This refers to the brain's activation in response to stimuli – whether this is fast and focused (therefore using less energy), or slow (which uses up more energy due to the young brain having to work harder to process and embed the information).

> Neural efficiency is not observable with behavioural approaches and one promise of brain rhythms is that they provide the opportunity to compare the higher-level processes that likely underlie humans' neural plasticity for language early in development in typical children as well as in children at risk for autism spectrum disorder.

The long-standing issue of correlation not being causal has long pervaded early childhood discourse, but brain studies such as those conducted by Bosseler et al. (2010), Percaccio et al. (2010) and Kuhl (2010) could signify the end of such challenges in applying findings from neuroscience to early childhood practice. Continual improvements in the techniques used will also contribute to the research being more useful in practical terms. Findings can also have implications for planning learning experiences that are truly inclusive of all children and their style and pace of information processing, problem-solving ability and concentration during tasks.

Brain imaging studies *are* significantly adding to our knowledge of phonetic development and learning (Meadows 2016; Kuhl et al. 2006; Werker and Curtin 2005; Kuhl 2004), and researchers hope that with greater precision of techniques, evidence will be used to progress understanding and inform practice. This might include determining whether babies are born with the innate capacities to learn a language (or whether these are switched on and developed as a result of listening to language). Neuroscientists hope they will soon be able to differentiate between brain functions unique to language and those shared with related cognitive functions such as speech pattern recognition, attention and working memory (the ability to mentally manipulate information and solve problems during short-term tasks). The implications for early childhood practice are also becoming clear. Part of Kuhl's (2011: 140) recommendations is the need to better understand the neural mechanisms underlying language development in order to provide excellent learning opportunities that are tailored to the needs of young children's brain development during the early period of life:

> The data on language and literacy indicate a potent and necessary role for ample early experience, in social settings, in which complex language is

used to encourage children to express themselves and explore the world of books. Further data will refine these conclusions and, hopefully, allow us to develop concrete recommendations that will enhance the probability that all children the world over maximize their brain development and learning.

It is fast becoming clear that neuroscience is opening doors to the mechanisms of the young brain with findings that are successfully demonstrating how the brain responds to and functions in response to the human voice. Kuhl (2010: 15) is optimistic about the contribution of neuroscience in helping us to continue unravelling the mysteries of the young brain and mind:

> Neuroscience studies over the next decade will advance our understanding of training methods for children with developmental disabilities struggling to learn their first language. These advances will promote the science of learning in the domain of language and potentially, shed light on human learning mechanisms more generally.

Pause for thought

1 Neuroscience shows that babies demonstrate improved speech perception and concentration as a result of listening to motherese. Discuss two practical implications of this when interacting with babies.

2 A parent asks you for some practical advice concerning how to further promote her nine-month-old daughter's efforts at communicating. How would you explain the benefits of the following?
- Joint attention
- Turn taking
- Sharing books

3 Summarize what fNIRS involves.

4 fNIRS identifies the neural processes that are involved in face-to-face interactions between infants and adults. How can this information be useful to practitioners who work with babies and young children?

It could be said that we are finally entering a golden age of neuroscience – where its applications are being used to help answer challenging questions that pervade aspects of early childhood development (Klin et al. 2015). The fNIRS breakthrough is just one exciting element of this, enabling neuroscientists and researchers to clarify what happens in the brain circuitry of children who have autism. The face-to-face conversations during fNIRS experiments demonstrate which brain regions become activated as well as those that fail to do so, as is the case with children who have autism (Redcay

2008). Brain imaging studies have also shown reduced activity and volume of the amygdalae in children who have autism (Abell et al. 1999; Bauman and Kemper 1994) and reduced connectivity between brain regions that play critical roles in emotional and social development.

Evidence provided by neuroscience – a help or hindrance?

Neuroscience is rapidly advancing in creating a revised understanding of cognition and language processing – and the neural architecture underlying language development. This includes revised interpretations of language processing, resulting in improved recommendations for intervention. Dr Jonathan O'Muircheartaigh of King's College London (2013: 16176) affirms:

> Our work seems to indicate that brain circuits associated with language are more flexible before the age of 4; early intervention for children with delayed language attainment should be initiated before this critical age.

O'Muircheartaigh's work highlights the need for interdisciplinary partnerships in order to ensure that children receive timely diagnosis, referral and provision of individualized services. While this does not mean that early childhood practitioners should be diagnosing autism, evidence from brain imaging studies can be used to support the interpretation of children's behaviour – and how best to support their communication and understanding of their world. Given that some experts now believe that autism can be evident in babies as young as two months (Jones et al. 2008; Grice et al. 2005), issues around appropriate brain imaging techniques and swift, sensitive intervention become all the more pressing (Klin et al. 2015).

The growing body of work which demonstrates the connection between neural circuits and language difficulties disprove the claim made by Bishop (2013: 3) that neuroscience has no place in the identification and intervention of language impairments in children.

> In no case has neuroimaging influenced the nature or application of intervention for children's language disorders.

As with many of the more recent disciplines, monitoring, rigorous assessment and time for consolidation are necessary, but these factors need not preclude the utilization of neuroscience-informed interventions. Until neuroscience becomes embedded in the identification of and intervention in language difficulties in young children, issues concerning accuracy of the methods used and the reliability of findings will continue to pose problems in its acceptance as a valuable method of intervention (Colheart and McArthur 2012).

Making recommendations that help bridge professionals' work at the practical level is therefore logical, but this needs to happen as early as is ethically and financially feasible. Correct diagnosis and interpretation of evidence arising from brain imaging studies is therefore critical. Although neuroscience is being successfully utilized in the identification of language impairments, there is still work to be done in order to improve the validity of findings from particular types of brain studies and how the findings are utilized. Goswami (2006: 24) proposes recommendations to address this current issue:

- To do far more methodological work to ensure neuroimaging tools are as reliable, sensitive and standardized as behavioural measures.
- To develop multicentre collaborations in order to do studies with adequate statistical power to detect treatment effects.

In conclusion, methodologies used in brain imaging studies (and interpretation of their results) do require further 'fine-tuning' in order to guarantee valid results that can assist researchers, medical professionals and early childhood practitioners in the selection of appropriate interventions. This will ultimately require changes to be made to early childhood, education and care policy at a national level to gradually incorporate findings from brain imaging studies that are applicable in early childhood practice (Oates et al. 2012). Effective health screening programmes and multidisciplinary working will be key components in this process.

Bilingual development

Infants are highly adept at learning a second language (or more), but that ability begins to fade rapidly – as early as their first birthday (Kuhl 2011; 2010). This echoes Pinker's belief that language acquisition is instinctive (1995). Just as monolingual infants' first exposure to language occurs in utero, so too does that of bilingual infants. Listening to sounds such as their mother's voice (Byers-Heinlein et al. 2010; DeCasper and Spence 1986) helps to build recognition of the native language from birth. Research has also shown that newborns are able to discriminate between differences in the sound of different languages (Byers-Heinlein et al. 2010; Mehler et al. 1988). This may well be linked to their 'in utero' experience of listening to their mother's voice. However, simultaneous and continuous exposure to both languages is required in order to help embed and develop the child's emergent languages. Observing both languages in action, listening, building understanding of each language's complex rules and features – and knowing when to use each language – is a huge task for the infant. It is little wonder that neuroscientists are grappling with this fascinating subject (Byers-Heinlein and Lew-Williams 2013; Poulin-Dubois et al. 2011). Byers-Heinlein et al. (2010: 343) describe the formidable task of understanding and speaking in two languages:

> In bilingual acquisition, infants must simultaneously detect and learn the regularities of each of two languages. This requires recognizing both languages as native while continuing to discriminate them.

Some researchers have shown that this sound discrimination begins with discriminating between the different rhythms or sounds of the languages spoken, with babies as young as six months tuning in to their native language (Kuhl et al. 2006; Werker and Tees 1984). It stands to reason that bilingual language development is a very interesting subject worth exploring, especially for early childhood students and practitioners. Understanding how to support an infant who is being raised to speak more than one language requires consideration and joint effort among staff in the setting. This will help ensure that the child is supported in line with their unique characteristics and that their parents can contribute their thoughts and ideas. This can be useful in the planning of experiences that is inclusive of their child's needs, which will rapidly change as their competence and confidence in using both languages develops. This requires understanding bilingual language development and how, in fact, it can enrich an infant's cognitive abilities and learning (Poulin-Dubois et al. 2011; Bialystok 2001; Yelland et al. 1993). In one of their research studies, Poulin-Dubois et al. (2011: 1) demonstrated that bilingual two-year-old children outperformed their multilingual peers in a set of tasks measuring executive functioning skills:

> Bilingual children have been shown to outperform monolingual children on tasks measuring executive functioning skills. This advantage is usually attributed to bilinguals' extensive practice in exercising selective attention and cognitive flexibility during language use because both languages are active when one of them is being used.

A few common questions asked by some parents and early childhood practitioners focus on the subject of confusion and silent (or non-verbal) periods: *Will growing up bilingual confuse the child? Why do some bilingual children go through silent periods?* Let us start with the idea that some bilingual children can get confused while acquiring two or more languages. What this looks like in reality is the child commonly switching between the two languages – sometimes during one conversation. This is an expected part of the bilingual child's language development (Pearson 2008; Lanza 2004) as they try to find the most appropriate word, particularly when they interact with parents who respectively speak their native languages with their child. It is also a great skill for the child to accomplish, given the attention, comprehension and expression of language involved. Another significant reason for this code-switching is usually because bilingual parents do so (Comeau et al. 2003). Some bilingual parents will code-switch with other bilingual speakers such as relatives and friends, so it is quite expected for their children to do likewise.

The silent period is another commonly experienced part of their bilingual child's journey. What this typically looks like is a period of time (which varies for each child) during which they do not communicate verbally, but might instead use hand gestures such as waving 'Hello' or pointing to an object they want. The silent period can be a deeply disconcerting and upsetting time for bilingual children, as they try to understand the norms of a new environment that is unfamiliar – both culturally and linguistically. It is an expected part of acquiring an additional language and is partly due to being away from the family home and the realization that their native language will not help them get by (for example) in a nursery. During this time, the child will actually be developing a repertoire of essential skills, such as familiarizing themselves with the social norms of the setting. This can include observing how the other children and adults interact, how children initiate play with their peers and gradually building an understanding of how everyone is expected to get along. Interestingly, studies (Kovács 2009; Bialystok and Senman 2004; Goetz 2003) have shown that bilingual preschool children show greater sensitivity to and awareness of others' ideas than monolingual preschool children.

Why does investigating infant bilingual language development matter? The more that evidence from neuroscience can be utilized in understanding how infants acquire two languages simultaneously, the better equipped practitioners will be when sharing science-based information with parents and primary carers concerning bilingualism, as well as using it to further enhance their work with young bilingual children. Poulin-Dubois et al. (2011: 578) inform us of the practical implications of their research, which has shown the advantages of bilingualism in infant executive functioning:

> At a more practical level, given the centrality of executive processes in cognitive life, the remarkable precocity of the bilingual advantage on these processes has significant implications for parents and educators who might be concerned with the effect of exposing their children to a second language early in life.

This should put to rest some parental and practitioner questions regarding diminished learning ability while acquiring two languages. What is important, however, is that those working with young children make the effort to understand the process of bilingual language development and view it as an advantage when it comes to learning. As a mother of a bilingual four-year-old child, I believe that raising her to speak Turkish and

English has only enhanced her development and propensity to learn. From pregnancy, I spoke and sang to her only in Turkish, while her father spoke to her in English. That remains the case today. The concept of one person, one language (Barron-Hauwaert 2004) does not work for everyone, but it remains a popular method for some parents (Byers-Heinlein and Lew-Williams 2013) and it worked for my family. Today my daughter effortlessly code-switches between the two languages without any hesitation. I strongly feel that encouraging her native language has also helped her to embrace its cultural values and traditions. It has also shown me that it takes consistent effort and determination to preserve one's native language, especially with reduced family networks to help reinforce it.

The tips for practice below might help you reflect on how you work with young children whose language and social skills need additional support in order to be effective communicators.

Case study   **Delilah's first word at eleven months old**

As a baby, Delilah was very talkative and curious – she loved communicating with anyone who would listen and show interest in whatever she was interested in at that moment. I always encouraged her attempts to communicate by talking to her (only in Turkish) about her experiences – like a football commentator guiding viewers through the match. Her favourite objects were two wooden crows. She would often play with these and I would always say the word for crow (karga) each time she picked them up. One day she suddenly announced to us 'karga!' It was her first word, and an unexpected one!

Top tips for practice!

- Be persistent in including bilingual children in small groups with other children.
- Provide activities (like role play) which encourage language practice.
- Ensure you have a cohesive team approach in supporting bilingual children's language development during the silent period.
- Give praise for the smallest effort!
- Encourage use of non-verbal responses (and Makaton if used in your setting).
- Observe the young child and talk about what she's paying attention to at that moment.
- Provide lots of opportunities during the course of the day for the child to engage in conversations with staff.

- Give the child visual cues including gestures such as moving objects within her line of vision to assure that she knows what you are talking about.

- Where possible, build positive relationships with the child's family. This can help when making suggestions to the family concerning what else can be done to promote their child's communication and comprehension.

- Always communicate clearly and give the child time to process information.

- Ensure the child has support in the playground, including the buddy system and circle of friends.

- For information concerning how to support children's language and creativity and reading and writing development, visit http://www.talk4meaning.co.uk/. This website is full of practical materials, including articles on emergent reading, spelling and writing and how to nurture young children's creativity.

Sustained shared thinking (SST)

Imagine that you are in the home corner at nursery and a boy (aged three years) invites you to play by picking up two toy mobile phones and passing one of them to you. You immediately respond to this spontaneous invitation by taking the mobile phone and having a conversation, and so a journey of sustained shared thinking begins for you and the child ...

You: Who should I call?
Artie: Erm, I know – call me!
You: Ok, that's a good idea. Ring ring, ring ring ...
Artie: Hi! (*Laughs*)
You: Hi Artie! What do you think we should do next?
Artie: Let's cook!
You: Ok, that's a good idea. What shall we cook?
Artie: Erm, sausages and sweetcorn.
You: That sounds good! Ok, see you later.
Artie: Bye!
You watch Artie put his phone down on the table and you do the same. Artie hands you a saucepan and a tin of sweetcorn.
You: What would you like me to do now, Artie?
Artie: If you cook the sweetcorn, I'll put the sausages in the oven.
You: That's a good plan! How long until the sausages are done?
Artie: Five minutes! I'll go and get plates!

You: How many plates do we need, Artie?

Artie: I'm getting two plates – one for me and one for you. (*Artie walks over to a cupboard and gets two plates. He brings them to the table and starts to dish up the food.*) That's for you!

You: Oooh thank you, it looks yummy! I think you've forgotten my sweetcorn!

Artie: Oh yeah! Here you go. Be careful, it's hot!

Siraj-Blatchford et al. (2002: 8) describe sustained shared thinking as:

> [a]n episode in which two or more individuals 'work together' in an intellectual way to solve a problem, clarify a concept, evaluate activities and extend a narrative. Both parties must contribute to the thinking and it must develop and extend the understanding.

This unlocking of potential might not seem revolutionary given Vygotsky's (1978) Zone of Proximal Development (ZPD) and Bruner's theory of scaffolding (1996). When we reflect on the influence of these theories on the development of sustained shared thinking, perhaps the key difference with SST is that *the adult engages in shared thinking with the child.* When this is done effectively, the adult provides just enough challenge to develop the child's understanding and competence in a given task. Helping very young children learn through active learning such as this can create neural networks in the orbitofrontal cortex (situated just behind the eyes and responsible for decision-making and problem-solving). Repetition of experiences that require children to be active and engaged in the learning process can lead to the strengthening of synaptic connections in the orbitofrontal cortex. Put simply, this means that the more involved a child is in their learning, the more beneficial it is for their developing brain (Stalnaker et al. 2015; Rolls 2004). Neuroscience is now providing further evidence when it comes to understanding just how much the young brain is capable of. Gopnik et al. (1999: Preface) explain:

> The research shows that babies and young children know and learn more than we could ever have imagined. They think, draw conclusions, make predictions, look for explanations and even do experiments. Scientists and children belong together because they are the best learners in the universe.

When babies and young children are encouraged to be curious, to explore and question their world, they can develop problem-solving skills, which can be beneficial later on life. Consider the young children in your work setting – each child will greatly differ given all that they have to contend with psychologically and emotionally. This puts the onus on practitioners to be adept at identifying each child's ability in light of their emotional state and provide support in order to help them achieve their full potential. A key part of this means working closely with parents or primary carers in order to develop strategies that take into account the affective factors that influence the child's reactions to the learning process.

Pause for thought

1 In what ways do you and your colleagues respond to young children's invitations to conversations?
2 What strategies do you employ to support young children to reflect on their learning?
3 Discuss one way in which sustained shared thinking might support brain development.
4 What are the strengths of sustained shared thinking in your setting?
5 In what ways can it be developed?

Top tips for practice!

- Ask open-ended questions which encourage more than one answer. This can help to develop young children's ability to think deeply, speculate and reflect. A few examples include:
  - 'What do you think might happen next?'
  - 'How did you ...?'
  - 'How could we find out more?'
- Do not focus on getting the correct answer, as this can inhibit young children exploring possibilities.
- Pay attention to what the child is saying and their body language. These will give you the biggest clues in helping you to frame your questions.
- Be spontaneous and unself-conscious! – Let the children lead you.

Below are the three broad themes mentioned at the start of this chapter. They are not intended to provide great detail but serve as starting points for you to reflect in relation to your experiences and to follow up those references that are of particular interest.

The influence of the home learning environment

The well-known saying 'home is where the heart is' is nowhere more aptly used than when discussing the importance of the home and all it represents for a young child. The home learning environment (HLE) and quality of parental engagement remain topical issues within the early childhood sector due to the knowledge we have concerning the impact that this has on an infant's all-round development and well-being – both in the short and long term. Although this section focuses on the role of the HLE with regard to children's language and communication development, the benefits of a good quality HLE extends beyond this. It can promote a child's resilience, general well-being and social skills (Sroufe et al. 2005). This is highlighted in the *Early Home Learning Matters* brief guide for practitioners (Family and Parenting Institute 2009: iv), which explains:

> How parents relate to their children from the moment of birth and the activities they do with them inside and outside the home during their early years has a major impact on children's later social, emotional and intellectual development.

There is no one definitive way to provide an effective HLE, but it should include some or all of the following elements on a regular basis: affectionate interactions, sharing books regularly, reciting nursery rhymes, playing games, providing a range of resources and activities that are appropriate to the child's age and stage of development, as well as trips to local parks, art galleries, museums, theatres, libraries and play groups. These experiences help instil a sense of curiosity and knowledge about the world and their place in it. When parents are their children's companions and can enjoy everyday life together with lots of interaction and talk, better developmental outcomes are the result for the child (Paterson 2011; Daycare Trust 2010; Sylva et al. 2004). Brain and language development are inextricably linked (Gopnik 2009, Bishop 2000; Eliot 1999). From birth, the experiences provided within the infant's environment result in the brain becoming hard-wired to learn language. Without adequate stimulation, however, a child may find it difficult to build a foundation for language development. The fundamental importance of a child's HLE and quality of experiences with regard to language development is highlighted by Kuhl (2011: 128):

> By the age of 5, prior to formal schooling, brain activation in brain areas related to language and literacy are strongly correlated with the socio-economic status (SES) of the children's families. The implication of these findings is that children's learning trajectories regarding language are influenced by their experiences well before the start of school.

It is therefore important that parents are supported to understand how they can provide an environment that is conducive to learning. As Bishop (2000: 138) explains:

In terms of brain development, the role of environmental input tends to be underestimated.

The quality of the HLE provided by parents also differs depending on a range of factors. These include:

- the type of attachment between mother and baby.
- emotional and mental well-being of the mother or primary carer(s).
- level of education of the parents/primary carers.
- family income.
- resources available to the child.
- support available to the family.

Evidence shows that good quality parental engagement which nurtures their child's innate curiosity and drive to explore and learn is vital in supporting their child's communication and language development (Gutman and Feinstein 2007; Sylva et al. 2004; Desforges and Abouchaar 2003). When children are supported in their efforts to interact with others, the outcomes can be very positive. Providing extra time, practice and predictability in the daily routine can help to build confidence. Specific activities and games can also encourage children to listen, speak and learn how to take turns during conversations. Making up and telling stories, identifying, describing and naming games are just a few examples of how this can be achieved. There is a plethora of research studies that explore the various speech and language difficulties experienced by children in their early years. The current emphasis on this may well be due to the sheer number of children presenting communication difficulties that require long-term intervention, as well as those research findings which tell us what happens when it's 'too late' (Marmot Review 2010; Allen 2011; Clark and Dugdale 2008). This current emphasis exists in various contexts, including the government's roll-out of free child care provision for two-year-olds living in disadvantaged circumstances, the review of the EYFS and early intervention strategies (DfES 2010). One in ten children have a communication difficulty in the United Kingdom, which requires ongoing speech and language therapy (ICAN, *The Cost to the Nation* 2006). Given that as many as half of all children start primary school with delayed speech and language skills (DCSF, *Bercow Report* 2008), the situation is clearly critical. Practitioners who can support parents to create daily opportunities to use and teach language and communication can help to meet this challenge.

Developing language in a digital age

It is a familiar scene: a family sits around the dinner table, each in deep concentration, attentively watching and listening to the voices in front of them. This all seems perfectly fine and beneficial for the young children who are also deeply engrossed in the exchange. The only issue is that they are not communicating with one another but are on their individual electronic devices. It has become all too common for some families to spend time together without really communicating with each other. Young children play hand-held computer games and visit social media websites, while primary carers spend time online on their mobile phones. Given that millions of people depend on the online world to fulfil various aspects of their lives, it comes as little surprise that excessive Internet use, in its many different guises, can cause long-term stress within families (Livingstone et al. 2011; Young 1998).

You may respond 'So what?' Sixty years ago, the arrival of televisions in many homes caused the same furore and yet no tangible 'damage' has been caused to families. The twenty-first century has seen a meteoric rise in the use of the Internet, the plethora of mobile phone 'apps' (applications), online gaming and social media. Research shows that children as young as two years old have regular, often unsupervised access (Holloway et al. 2013; Livingstone et al. 2011). The last six years has seen a substantial rise in children aged below nine years using the Internet (Helsper et al. 2013). Concerns regarding early brain function and the ability to focus have also been recognized (Christakis 2009; Zimmerman et al. 2007). Research studies on infant brain development and language acquisition (Kuhl 2010; Hayne et al. 2003; Barr and Hayne 1999; Meltzoff 1988) consistently show that in order to acquire language, it is not enough for infants to passively watch somebody talking on a screen. Meaningful interactions in 'real time' are vital in facilitating language and communication.

The influence of the digital age is now a core part of many home learning environments for millions of families. The ubiquity of the Internet and mobile phones and the multitude of 'apps' and tablet computers is resulting in children competing for their parents' attention – and vice versa. A survey for the charity Action for Children found almost one in four mothers and fathers (23.1 per cent) struggle to control their children's screen use. Given that some of the most popular electronic devices are now hand-held, parental supervision has also become more difficult. The advent of such small screens also makes sharing of the experience and talking about it less easy to achieve. In short, these electronic devices can cause children to isolate themselves and, as a consequence, frequently miss opportunities to engage with their family.

One report published by Ofcom UK (2015) indicated that at least three in four children live in a household with a tablet computer, with 53 per cent of children aged between three and four years regularly using a tablet. The report also shows that 77 per cent of children aged between three and four years regularly use a desktop computer/laptop/netbook with Internet access. The early years of a child's life are crucial in terms of laying the foundations for intellectual, language, emotional and social and physical development. Parents, primary carers and early childhood practitioners therefore have a moral responsibility to ensure that the use of ICT devices enhances these vital life skills. Vygotsky's concept of cultural tools for learning still plays an integral part in early years curricula. His emphasis on the social context and the use of cultural tools to enrich learning is evident in the Early Years Foundation Stage (EYFS), which takes into account the constant presence of information and communications technology (ICT) in our lives and guides practitioners how to incorporate its use in meaningful ways. In the spirit of Vygotsky's philosophy, the EYFS is designed to encourage the use of ICT across the indoor and outdoor learning environments through provision of tools for discussions, collaborative problem-solving and online support systems to scaffold children's evolving language, understanding and intellectual development.

Where activities with ICT devices are educational and supervised, the outcomes can be positive for very young children. As mentioned, early childhood curricula make it compulsory to embed ICT across all subjects taught, in recognition of the necessity of ICT in our world. This makes sense given that technology permeates almost every aspect of our lives, which makes the issue of proficiency all the more important, so that children do not get left behind in this digital age. Having opportunities to use ICT in the real world is all the more pressing for children whose backgrounds are financially disadvantaged, as they may lack the opportunity to use ICT (Thouvenelle et al. 1994). Embedding ICT across all areas of early years curricula is just one effective way of helping to ensure that no child is left behind. Below are just a few practical suggestions that may help to make ICT work in your setting.

Pause for thought

1   Consider the young children in your setting and their use of ICT in their HLE.

1a Are you aware of the amount of time spent on devices such as tablets and online computer games in the HLE?

1b What practical use might this information have for early childhood practitioners?

1c What steps has your setting taken to inform primary carers about the advantages and potential limitations of ICT use among very young children?

2   How far do you think it is your responsibility to discuss home ICT use with parents?

Top tips for practice!

- Make sure that computer (and other ICT) use is not just seen as a reward, but integrated across all areas of learning and development as part of your curriculum.

- Take ICT outdoors! Encourage children to use equipment such as cameras and video recorders in the garden to capture moments of personal interest. These can be used to extend learning in the classroom.

- Make sure that all computers are accessible and that all children know they can use them.

- Let the children be in control as far as possible, so that they can build problem-solving skills as they navigate their way around the ICT resources.

The role of intersubjectivity in nurturing early language and communication development

The role of early parent-baby interactions in the development of language and communication development

Intersubjectivity may not be a term that is readily used in your daily work with babies and young children, but it is a core element of respectful and responsive relationships – the three Rs if you will. Intersubjectivity can be defined as the close relationship between two individuals that is characterized by joint attention, in which the movements, non-verbal gestures like eye contact, hand gestures and facial expressions and language are in complete harmony. Hughes and Baylin (2012: 103) define intersubjectivity as:

> [t]he reciprocal relationship between a parent and child. Within this mind-to-mind rapport, both the parent and child re-open to one another, receptive to and sharing each other's experience of self, other and the world.

From birth, babies are highly perceptive and emotionally aware. They can recognize and are comforted by their mother's scent, her voice and her touch. Babies, then, are by nature primed for affectionate care and stimulation to help ensure healthy brain development even from before birth (DeCasper and

Prescott 2009; DeCasper and Spence 2006). Trevarthen and Delafield-Butt (2013: 5) observe:

> A newborn infant's movements are especially sensitive to sight, hearing and touch of an attentive the mother in face-to-face engagement, and they can take a creative part in a shared narrative of expressive action.

Within hours of birth, babies demonstrate their readiness and ability to communicate (Trevarthen 2011). It is these life-affirming, pivotal interactions between the mother and her baby that shape the baby's ideas about their place in the world, how they expect to be responded to and how relationships work. Consistency of affection, shared eye contact and conversations in play and care-giving tasks all contribute in helping to build the baby's understanding of language and its uses. For example, these factors can enable the baby to start making associations between the words spoken and the objects being pointed to – and eventually learn their first words (Morales et al. 1998):

> When young children reach out for interaction, an adult who responds to the child helps to build and strengthen the child's brain development. They do this through creating a relationship where a child's experiences are affirmed, nurtured and supported.

In order for all professionals who work with and for children and their families to make findings from neuroscience more useful in nurturing early growth and development, good practice needs to be recorded and shared in order to build up evidence of what works and how. This documenting and monitoring will enable more evidence-based ways of working that can then be disseminated across settings and teams, with a view to better promoting language and communication development.

Pause for thought

1 How do you put the three Rs into action as part of your interactions with very young children?
2 Which aspects of your interactions do you think could be improved and why?
3 Describe the characteristics of intersubjectivity.
4 Consider the provision in your setting for promoting babies' and young children's language and communication development.
4a How do you currently share good practice with colleagues externally?
4b How might this be improved?

Top tips for practice!

- Hold regular staff meetings to share knowledge about the babies and children, in order to further enhance support for parents in promoting their baby's learning in the home.
- Support effective parenting by having the skills to offer evidence-based interventions.
- Ensure staff undertake regular training that is targeted on building knowledge concerning the role of healthy attachment in early brain development.
- Make language accessible for all children from the earliest age: give them choices during the daily routine such as snack time, play time and structured activities.

Autism

This section explores the neurodevelopmental disorder known as autism. At its core, autism is actually a neuropsychiatric condition as its social, emotional and cognitive symptoms occur as a result of malfunctions in the nervous system (Frith 2003). It is discussed in relation to recent research findings from neuroscience, which are enabling scientists, psychologists and educational professionals to distinguish the source of its many complex characteristics at an earlier stage than has previously been possible. The signs and severity of autism vary for each child: some will show few signs of the disability while others will experience more of the associated characteristics. Children who have autism can find it difficult to form friendships, preferring to play alone, often repetitively. Lack of imagination and repetitive body movements are also common, such as flapping hands and spinning. They may also have difficulty in understanding others' thoughts and feelings, which is known as Theory of Mind (ToM). ToM concerns awareness of and the ability to understand others' mental states based on their behaviour. Happe and Frith (1999: 1) define ToM as '[t]he ability to attribute mental states and predict behaviour accordingly'.

Having ToM is fundamental to successful social interactions, helping children to understand others' behaviour and verbal/non-verbal language and in adapting their responses to them based on their behaviour. Usually, a child with ASD does not have ToM and can hence find it challenging to understand others and get along at nursery and school. Children with ASD tend to have delayed language development as a result of their delayed communication and language development, which is characteristic of ASD. All these signs can be categorized under three broad headings, as identified by Baron-Cohen (2004: 1):

Autism is diagnosed when a child or adult has abnormalities in a 'triad' of behavioural domains: social development, communication, and repetitive behaviour/obsessive interests.

We know that young children enjoy the attention of a familiar adult, but children who have autism tend to find it difficult to initiate joint attention with an adult and can feel uncomfortable when interacting with others. For example, if a child cannot distinguish between facial expressions, they will find it very difficult to determine the appropriate emotional or social response, especially since what we say often does not match what we do (such as body language and facial expressions).

Autism affects approximately 700,000 people in the UK (National Autistic Society 2016) and approximately 1 in 68 children in the USA – but this figure does differ across each state (Centres for Disease Control and Prevention 2016). The warning signs that primary carers and early childhood practitioners should look out for if they are concerned that a baby or young child has autism include the following:

- not smiling back at carers or familiar faces by six months.
- not giving eye contact.
- not initiating or responding to cuddles.
- not babbling by twelve months.
- not using gestures to communicate or to point at objects.
- no spoken words at sixteen months.
- a lack of pretend play.
- inflexibility of thought, which can manifest as repeating the same movements and actions (e.g. tapping ears, lining up toys and flicking their fingers).

Recent evidence from neuroscience concerning the causes of autism shows their inception in pre-natal brain development (Trevarthen and Delafield-Butt 2013; Rodier and Arndt 2005; Prechtl 2001) and that errors in the brainstem in particular are a primary cause of a range of developmental difficulties post-natally. Trevarthen and Delafield-Butt (2013: 2) explain:

A primary cause of autism spectrum disorders is an error in early growth of intrinsic motive and motor systems of the brain-stem during prenatal ontogenesis.

Post-natally these errors manifest as poor sensory-motor development (including the receiving of sensory information from the body and the environment through the five senses) and poor motor coordination and sequencing of movements. It is partly due to this poor coordination and sequencing of movements that Trevarthen and Delafield-Butt (2013) propose that children who have autism generally experience language difficulty. Lashley (1951: 121–2) explains the close relationship between the ability to sequence and acquiring speech, in terms of the underlying make-up of the brain and mind:

> Not only speech, but all skilled acts seem to involve the same problems of serial ordering ... Analysis of the nervous mechanisms underlying order in the more primitive acts, may contribute ultimately to the resolution even of the physiology of logic.

More recently Trevarthen and Delafield-Butt (2013: 2) adopted a similar approach in researching the causes of autism at a neurobiological level. They believe that research such as theirs, which identifies the key, pre-natal factors that cause autism (such as faulty brainstem development and poor motor affective abilities), can lead to earlier diagnosis with multidisciplinary intervention being put in place for the child earlier:

> We propose, with evidence of the disturbances of posture, locomotion and prospective motor control in children with autism, as well as of their facial expression of interest and affect, and attention to other persons' expressions, that examination of the psychobiology of motor affective disorders, rather than later developing cognitive or linguistic ones, may facilitate early diagnosis.

Again, emphasis is put on *first* understanding the brain and mind, and the errors in brain function which result in autism (Frith 2003), before treating its secondary symptoms (such as the associated social, emotional, cognitive and linguistic difficulties). Findings from neuroscience could provide information that can help early childhood practitioners to better understand children with ASD on the spectrum, and can lead to innovations to work on areas of difficulty that are commonly experienced.

Pause for thought

1 How would you describe the connection between coordination and language development?
2 Suggest two practical ways you could support the:
   a) language development of a young child who has autism
   b) coordination/motor skills of a young child who has autism.
3 Identify two ways in which findings from neuroscience are contributing to the earlier detection of autism.

Receiving earlier support can make all the difference when a young child is experiencing difficulty in using language to express their thoughts and feelings, and when trying to make sense of their world emotionally and socially. Earlier diagnosis can also help parents to better understand what autism will mean for them and their child – and how best to build a bond and interact with their child. This will prove very important for families affected by autism because of the great difficulty experienced socially and emotionally. Warm and responsive communication can be protective as well as encourage the young child's attempts at socialization (Nagy 2011; Sander 2008; Brazelton and Nugent 1995) without pressure being unintentionally placed on the child. We often make the mistake of trying to encourage children with autism to occupy 'our' world; perhaps we should be trying to occupy their world.

Case study

Below is a case study written by a mother who has three children, aged seventeen, eight and ten years old. One of her sons, Luke (aged ten years old), was diagnosed with autism a few weeks before his fifth birthday. Read the case study and reflect on the questions which follow.

Luke's diagnosis was made early, before he was five. He was not given a statement of needs until Year 1 in January, just before he was six years old. My concerns for Luke began when I identified that he was failing to reach all his physical milestones like sitting, crawling and walking, which were all delayed. His speech was also delayed and it was only after a series of hearing tests at hospital that fluid in the ear was identified, which had probably been present from an early age. (This cleared by itself when he was four and he did not need grommets.) As a result, his speech is still unclear, but this is also linked to his cleft palate which has been repaired twice. Luke displays many of the traits associated with ASD, such as flapping his hands, repetitive interests, and patterns of behaviour. He also shows an interest in the sensory aspect of play, linking with his obsession with soft toys, real dogs and cats.

In addition, Luke has global developmental delay. As a result, his needs are complex and at times frustrating as we try to understand how his development is affected by which symptoms. Supporting Luke has meant that he needs one-to-one supervision and regular meetings with a host of professionals which mainly come under the hospital's cleft palate team. Annual statement reviews take place at his school and other than this there is not much else in terms of guidance that we as parents are given. His reviews indicate progress (progress is, however, slow: to date he is not reading or writing) and he is moving onto the next stage in his development. My partner and I read around the subject of autism as it is not clear to us what the biggest hindrance to his development is. To be honest we are just trying to manage and understand him as best we can so he grows up happy and secure. This is very difficult as we do not always know what we are doing or whether the decisions we are making are the best choices for him in the future.

We have found that the only way to motivate Luke is to build on what he is interested in. At the moment this is pirates, Teletubbies (the yellow one mostly) and people he knows and who are important to him. We find that Luke has an obsession with the iPad and will use it to feed his obsession with particular interests. Over Christmas he would watch the same clip from a CBeebies programme over and over again. However, the televised world seems to have advantages and disadvantages. One disadvantage is that increased time on the iPad affects behaviour – he can often be aggressive and irritable. We are constantly trying to get the balance right for Luke.

Pause for thought

1 Discuss two practical ways you and your colleagues could support a child with ASD, in order to ensure that they can access all the curriculum has to offer.
2 In what ways could Luke's interests be incorporated into the setting's planning of activities?
3 How would you work with Luke's parents to best nurture his socio-emotional skills?

Top tips for practice!

● Ensure the child has a predictable routine as this can help them to make sense of the day and can increase their confidence.
● Encourage the child to get involved in conversation and play with other children while being patient as they try to understand the 'rules' of communicating (listening, taking turns to speak and waiting as others talk).
● Make regular links to things/people that are important to the child.
● Make time for lots of singing and rhyming as these can help to build the child's phonological awareness.

Concluding thoughts

This chapter has looked at the connection between neuroscience and its role in the identification of language and communication difficulties experienced by some children. While neuroscience still has some way to go in being fully accepted and utilized by practitioners and professionals concerned with early language development, it continues to rapidly become more familiar as part of the early childhood discourse. Some of the more commonly encountered language difficulties and factors that impact on young children's language development have been discussed, highlighting how neuroscience is helping to build an understanding of these. Hopefully this knowledge can lead to more evidence-based programmes of intervention for children who have atypical communication and language development and their associated behaviours. Increased collaboration between neuroscience, early education and psychology can also result from findings that have practical implications for early childhood professionals. As a result, we hope that they can be better equipped to offer parents and primary carers strategies that could help them to better understand the impact of neurodevelopmental disorders and how they can work together to put into place early, intensive support that promotes the social, emotional, cognitive and linguistic development of children.

Further reading

Frith, U. (2003). *Autism. Explaining the Enigma.* Oxford: Blackwell.

This engaging book discusses autism, with links made to case studies throughout history to help the reader understand what autism looks like and feels like for young children who live with the condition. The history of autism is brought up to date with an examination of the application of neuroscience in the diagnosis of autism and how this has led to increased and earlier diagnosis of the condition. The author also examines genetic research and the part it plays in identifying 'faulty' genes which can cause autism. This stimulating read is a useful resource to dip in and out of depending on your particular area of interest.

Kuhl, P. (2010). 'Brain Mechanisms in Early Language Acquisition', *Neuron* 67 (5): 713–27.

This is a highly informative research article concerning language acquisition and brain development from birth. Professor Kuhl engages the reader as she identifies the neural networks that constitute the 'social brain' and its role in early childhood language acquisition. Kuhl also discusses how neuroscience studies using Magnetoencephalography (MEG) can assist in identifying the differences in brain responses to speech in infants who have autism spectrum disorder, and those who do not. Commonly used techniques are examined, with a balanced look at their benefits and limitations. The paper concludes optimistically regarding the role of neuroscience in leading theory concerning typical and atypical language development.

Sprenger, M. B. (2008). *The Developing Brain: Birth to Age Eight.* Thousand Oaks, CA: Corwin Press.

This is an easily accessible book that is full of practically useful information for those who care for and work with babies and young children. The step-by-step guide is easy to understand and apply. The eight chapters take us through a child's all-round development, with an emphasis on brain development and key concepts (such as windows of opportunity, neurotransmitters and experience-dependent and experience-expectant learning) at every developmental stage. This book is a good quick reference resource for those who are new to learning about the developing brain.

Chapter 5
Beyond Nature versus Nurture: Is Neuroscience Relevant to the Debate?

What this chapter is about

There are moments where we observe the behaviour of our own children or that of others in the work setting, and we wonder 'Where does she get that trait from?' or even 'How can twins have such different temperaments?' This chapter introduces and unpacks some of the current issues concerning the age-old nature versus nurture debate in light of evidence provided from neuroscience. Factors that shape pre- and post-natal human development and related concepts such as epigenetics and how gene expression can become 'set' during the pre-natal period are included. Woven into these discussions is the importance of understanding brain development as key to quality, with emphasis on the significance of secure attachments and nurturing environments. Consideration is also given to the influential role of early childhood practitioners and primary carers, alongside applicable theories to help contextualize these ideas within early childhood education and care.

Why you should read this chapter

Anyone with an interest in child development and its role in helping to improve the early experiences and life chances for babies and children will benefit from reading this chapter. It is an easily accessible introduction to topical issues concerning child development through the lens of epigenetics and neuroscience. By reflecting on the case studies and questions, you will be able to consider the topics at a deeper level and challenge aspects which you feel do not correspond with your experiences. Thinking about these ideas and developing your own opinions can lead to a consideration of the 'bigger picture' when it comes to understanding the origins of behaviour and how to bring out the best in babies and children. The advancement of educational neuroscience and neuroimaging techniques now show the inextricable link between genetics and the environment. As Ridley (2011: 280) succinctly puts it:

> Nature versus nurture is dead. Long live nature via nurture.

The integration of neuroscience and genetics has positive implications for early childhood practitioners, health care professionals and policymakers concerning prenatal maternal health and the promotion of child development and learning. Together, these rapidly evolving disciplines provide insights into how we can play our part in changing the life trajectories of those children who might otherwise struggle to realize their full potential. Early intervention that is informed by neuroscience may well prove increasingly important in this process.

What is the nature versus nurture debate?

The answers to the mystery of human development in relation to the nature versus nurture debate vary enormously and depend on a range of influences. Nativists (supporters of the nature side of the argument) believe that aspects of human development such as intelligence and personality are determined by genetic make-up. Empiricists (supporters of the nurture side of the argument), however, believe that these are acquired (i.e. learned). Empiricists such as John Locke popularized the Latin phrase *tabula rasa* (meaning 'blank slate'), conveying the idea that the child's mind is a blank canvas on which the adult can inscribe knowledge and experience. Noam Chomsky's Language Acquisition Device (1965) is just one example of the nativist perspective of human development. He believed that the ability to understand and reproduce language was innate in all humans due to an existing mental capacity and the requisite vocal mechanisms. Theoretically, this of course makes perfect sense: a baby observes her parent and upon babbling or cooing back, they praise her and engage with her to further encourage her communication, and so the proto-conversations develop.

Pause for thought

1   Why do you think the nature versus nurture debate is part of the early childhood discourse?
2   Some people believe that the prenatal environment is nature, while others view it as nurture. What do you think and why?
3a Do you think that characteristics are mainly determined by nature or nurture?
3b Do you think it is possible to answer this question definitively? Explain the reason for your answer.
4   Do you think it is important to distinguish whether characteristics are determined by nature or nurture? Please explain your answer. Some people put pre-natal environment in nature but others count that as nurture.

Unfortunately, some children do not benefit from listening and responding to their primary carers' positive expression of language. Tragic examples of children raised with a complete lack of care and affection, with no (or only aggressive) verbal communication from their parents, include that of Oxana Malaya and Genie. Neglected and starved of affectionate human interaction, Oxana and Genie never developed language while living with their abusive and negligent parents. Both children did not learn how to stand, walk, talk or develop self-help skills. Furthermore, Genie was beaten each time she made a sound. Their only food source was provided by eating scraps of food or excrement. Although extreme, such an example is a tragic reminder of the unequivocal role of affirming adult input to nurture the healthy growth and development of children. We may be well versed concerning the factors that contribute to nurture, but it's *how* we use this knowledge to intervene and support those young children who most need it that is more important. John Locke's well-known paper, *An Essay Concerning Human Understanding* (1690/1722), provides evidence of the stark contrast with today's beliefs about the origins of knowledge and identity. According to Locke, babies entered the world completely blank and thus were lacking any innate behaviours, knowledge or understanding. Each of these were to be bestowed by parents, educators and other significant adults. In Book I of his essay, some of these beliefs are introduced early on, where Locke (1690/1722: 62) passionately argues against the idea of nativism:

> If we will attentively consider newborn children, we shall have little reason to think that they bring many ideas into the world with them.

On the contrary, when we do observe and consider the newborn child, their curiosity, aptitude for sociability and connection are clearly demonstrated through their movements and tracking of humans and objects which are in close proximity. Another notable empiricist, Albert Bandura (1961), believed that learning and development occurred as a result of observing others. His well-known Bobo doll experiment provides a good example of children learning and adopting aggressive behaviours through observing the actions of adults who also displayed aggressive behaviour towards the doll in the experiment. Many children are still allowed to play violent computer games and watch programmes that display aggressive behaviour. Unfortunately some also witness acts of domestic violence in the home, which can have devastating and long-term effects on their holistic well-being. (This will be discussed further on in this chapter.) The murder of two-year-old James Bulger (1993) at the hands of two children, both aged ten, is a chilling reminder of what can go terribly wrong when families do not provide the necessary security, stability, affection and reasonable boundaries that children require in order to thrive. Tales of parental neglect, alcohol abuse and violence were all cited as factors in the murderers' home backgrounds. Also of great significance were certain horror films that the boys watched repeatedly and emulated as part of their torturing and murder of James Bulger.

The following case study is provided by a nursery manager based in inner London. She describes how their provision helped to improve the all-round development of one infant (aged one year old) whose mother was experiencing domestic violence.

Case study

This case study tells of a twenty-month-old boy who joined our nursery shortly after his first birthday. The family are practising Muslims, and live in a housing estate close by to the setting. The child's first language is English although some Arabic is communicated between the parents at home. The nursery place was sought by Social Services using a funded (three hours daily) place to support the family. Information was shared that the mother had been a victim of domestic violence and although the relationship had ended, the father was now living back at the family home. It soon became apparent that while attending the settling-in sessions, the child's mother was very apprehensive about her child attending nursery and I felt the child attending nursery was due more to a recommendation from Social Services than a decision reached by the parents.

The child lacked confidence and his behaviour was timorous. He appeared unsure and withdrawn in comparison to his peers. Rather than play alongside his peers, he would rather stay close to his key person once his mother felt comfortable enough to leave him. He continually sought out a familiar one-to-one bond; this is where the role of the key person has worked wonderfully and has been a great emotional support for the parent and the child, providing both reassurance and comfort. We recognize that parents are the most important people in their children's lives – they know the most about their child and hence their partnership with the nursery team is significant to the child's attainment. During the settling-in sessions the child's key person and I explained the types of activities the child could participate in during his time with us, the relationships with other children he would form and that we wanted his mother to be involved in planning for her child – sharing his interests and what makes him smile and laugh.

Our policy on engaging parents has one clear purpose: to act as a vehicle for valuing and supporting parents to give their children the best start in life. We use a home learning approach to ensure we have regular pedagogical conversations that enrich the parental experience and clarify how we can work in partnership to improve the child's emotional, social and educational success.

We recognize the positive experiences that this child gets from attending nursery is just as important as avoiding those negative experiences he may have witnessed between his parents. The process of developing a strong sense of self is based on the quality of engagement that the child gains from his time at the nursery. The environment is designed so that children are encouraged to explore, investigate and be creative both independently and with others. As the child has grown in confidence and familiarity within the setting, he has become more involved in different learning experiences. Heuristic play was used to increase this child's confidence as there is no right and wrong in his exploration, so failure never occurred, and the child was able to learn to play alongside and cooperatively with his peers. This child has learned to notice his own feelings during his past eight months at the nursery. This is achieved through feedback often using gestures and single words from staff. For example, praising and encouraging him when participating in an action song or creating a piece of art. This has led to him repeating the action and feeling more comfortable with staff praising his efforts.

Having reached a level of self-awareness, he can use it to become aware of other people's feelings and build empathy towards others, for example stroking a younger baby's head when upset. Educational and all round development for this child may have been very different had he not experienced the nurture provided by us. There is no doubt that he was at risk of delayed speech, social withdrawal, delayed cognitive abilities and delayed physical motor skill development. I believe that having an understanding of how the brain develops and how this affects behaviour was significant in our ability to respond sensitively and appropriately to this child. It's an area that can be overlooked within a training programme but given the long-term impact poor brain development can have on a child's overall cognitive development, it is a subject that needs to be highlighted given the first three years of life is a period of incredible growth in all areas of a baby's development.

Pause for thought

1 In the case study, which three factors did you find the most protective of the mother and child?
2 Given that the domestic violence had been present prior to the child's birth, how might the domestic violence have affected his brain development pre-natally, given his temperament when he first joined the nursery?

To summarize, the role of genetics, or nature, cannot be underplayed or replaced by a completely empiricist perspective. After all, genes provide the blueprint for our existence. What we should take on board more readily is the highly complex interplay between nature and nurture in the rearing of healthy and well-rounded human beings. For example, parents and primary carers who take the time to talk openly about feelings and help their children to identify and express their feelings are more likely to raise them in ways that are more emotionally literate than those who do not invest time and effort doing so. Nurture therefore plays just as significant a role in facilitating healthy development as those driving innate factors. The nature versus nurture debate is no longer viable – neuroscience evidences the complex interplay between the two, with early intervention being all the more critical given that nature and nurture play such a significant part in determining educational outcomes and cognitive, language, behavioural and socio-emotional ability.

Where are we now? How epigenetics helps to solve the nature versus nurture question

One of the latest scientific breakthroughs comes in the form of **epigenetics**. This is the study of variations in **gene expression** that do not involve changes to the genetic code but still get passed down to at least one subsequent generation. So, it concerns reversible heritable changes that do not alter the DNA sequence, but do switch genes on or off due to chemical changes around the genes which occur over time. Gene expression is the process by which genes are either *silenced* (switched off) or *expressed* (switched on) in different tissues and cells during conception. Figure 5.1, below, depicts some of the positive and negative factors that can influence gene expression.

Figure 5.1 Epigenetics – factors which can influence gene expression

Epigenetics came into existence during the early part of the twentieth century (Waddington 1939), but it is in the past two decades that it has really gathered speed. Scientific research is currently focused on unravelling the epigenetic mechanisms related to the types of changes that take place in gene expression as well as discoveries concerning the relationship between changes in gene expression and the factors affecting this process, such as maternal physical and mental health, neurological and psychiatric disorders and immune disorders. It may not be immediately obvious, but epigenetics is part of our daily lives. It shows us that where we live, the foods we eat, the amount of physical activity we undertake, our sleep patterns, the presence or absence of alcohol or drugs, the environment, lifestyle and stress each play a role in influencing gene expression (Ridley 2011; Greenspan 1995). It is these factors that can cause the chemical changes around the genes and hence switch them on or off, over time. Historically, the discourse surrounding the nature versus nurture debate has been viewed in finite terms, demanding an 'either or' response. Crucially, epigenetics is helping to transform the discourse, pushing it beyond this limited perspective. It shows us that nurture influences nature – a concept previously thought of as fixed, decided by the powers that be, before birth. Psychologist Sara Meadows (2016: 57) explains epigenetics as being the result of the interaction between the individual and its environment:

> We have to consider development as involving adaptation between organism and environment and as absolutely crucial to the interplay between genes and environment, and we are talking about genes and their expression being affected by their development. Thus, we are talking about *epigenesis*.

How is epigenetics useful to you?

This exciting branch of science gives us evidence regarding the pivotal role of external factors that can influence gene expression and, ultimately, human development. As stated, factors such as lifestyle choices, the environment, the presence of prolonged stress and maternal health each play a role in determining gene expression and brain architecture. In your work with families and young children, being well versed in the factors that can influence gene expression and offering sensitive, timely and continued practical support can help prospective parents to think about their current lifestyles and make any beneficial changes. Providing practical advice to help families cope with any long-term stress or trauma, or obtaining early intervention during pregnancy, can help to minimize the possibility of detrimental gene activity from occurring in response to the environment.

The role of epigenetics and neuroscience in explaining early brain growth

Although the brain contains approximately 86 billion neurons at birth, the greater part of neural development actually occurs during the prenatal period and dies off soon after birth, which makes prenatal brain development a critical time. Factors such as poor nutrition, drug and/or alcohol abuse, disease, dependency on certain prescription drugs and chronic maternal stress can consequently impede brain growth and development even before the child is born. We therefore cannot underestimate the importance of the prenatal period in terms of its influence on brain development (Fox and Schonkoff 2011), as the two are inextricably linked. Rose and Abi-Rached (2013: 194) comment:

> It seems obvious that intensive intervention on early parenting, on the parents of today, and on those at risk of being the dysfunctional parents of tomorrow is the path required to break the cycle of antisocial and violent conduct that destroys lives and costs our societies so dearly.

Some of the most influential studies concerning epigenetics originate from neuroscience research. One example is Meaney's persuasive research study (2001), which demonstrated the influence of maternal behaviours on neural growth. Meaney's study showed that there is a direct link between maternal care and the handling of pups by rat mothers (licking and affection given to her pups), which leads to alteration in the sculpting of neural circuits especially in the sensitive period of plasticity. As Weaver et al. (2004: 847) explain, this affirmative maternal behaviour towards her pups exerted a positive influence on the pups' gene expression, which consequently improved their response to stress, which, they say:

> [...] altered the offspring DNA methylation patterns in the hippocampus, thus affecting the development of hypothalamic–pituitary–adrenal responses to stress through tissue-specific effects on gene expression.

Crucially, similar conclusions were reached in human studies carried out on post-mortem brain tissue of suicide victims who had a history of childhood abuse (McGowan et al. 2008, 2009). If you would like further details about this interesting study, please refer to the Bibliography, but the important take-home message from their study is that early life adversity does play a critical role in altering DNA methylation patterns and hence neural development, which leaves a legacy on the brain long into adulthood. McGowan et al. (2009: 346) determined that:

> Early life events can alter the epigenetic state of relevant genomic regions, the expression of which may contribute to individual differences in the risk for psychopathology.

The plethora of books, articles and reports concerning early brain development generally expound the importance of building a healthy brain from birth. Yet it is far more useful to start from conception. Professionals in the field now refer to this crucial developmental phase as the first 1001 days – the time from the start of pregnancy to a child's second birthday. This is due to research consistently showing the impact of children's early experiences on their adult emotional and mental health as well as their educational and employment opportunities (Paterson 2011; Barker 1995). The highly publicized *1001 days campaign* (WAVE Trust report 2014; United Nations Standing Committee on Nutrition 2010; Barker 1995) highlights the importance of the in utero environment with regard to growth and development, from conception to a child's second birthday. Evidence cited in the WAVE Trust report (2014: 5) further emphasizes the critical importance of the 1001 days in terms of influencing health and development in the short and long term:

> Ensuring that the brain achieves its optimum development and nurturing during this peak period of growth is therefore vitally important, and enables babies to achieve the best start in life. Whether out of concern for an individual baby's well-being or safety, or for the costs to society of poor attachment, it is imperative that how children are raised is guided and influenced by this principle and evidence.

Research also demonstrates that a poor maternal diet that is high in processed foods and lacking in protein and essential fatty acids, also known as EFAs (found in nuts, unprocessed oils and oily fish such as mackerel and salmon), can cause direct epigenetic effects in the foetus with resulting susceptibility to lifelong conditions such as diabetes, obesity, heart disease and reduced lifespan (Prado and Dewey 2014; Lillycrop, et al. 2009). Early brain development can also be affected by poor maternal diet as the myelination process becomes interrupted by a lack of sufficient nutrients that are required to form healthy neurons. The EFAs (which cannot be made by the body) provided by a healthy diet are actually structural constituents of the myelin sheath. Where EFAs are lacking in the diet, myelination does not occur effectively, resulting in a slower transmission of signals between neurons. Prado and Dewey (2014: 269) tell us that the optimum time frame for myelination to occur is up to the age of two years:

> Myelin is white, fatty matter that covers axons and accelerates the speed of nerve impulses traveling from one cell to another. The most significant period of myelination occurs from mid-gestation to age two years.

The first 1001 days are not only crucial for the baby but for the mother also. Parenthood is a time of immense upheaval and change, which can lead to feelings of intense anxiety, uncertainty and, in some women, depression. These feelings may manifest alongside existing issues such as lifestyle habits that are highly detrimental to the well-being of the foetus, well beyond childhood. The lifestyle habits or choices of a pregnant woman therefore have a huge impact on her baby's brain development (and all-round development). For example, when a parent chooses to breastfeed, the benefits for her baby include protection from infections and diseases while helping to build a strong physical and emotional bond between mother and baby. Not all mothers can (or choose to) breastfeed: this does not automatically mean that the bond between mother and baby will be compromised. It is, however, worthwhile noting that the constituents of breast milk cannot be identically matched. No formula milk can offer colostrum – the fatty 'first milk' – but feeding, no matter how, is a good time to strengthen that bond. Conversely, smoking, drinking heavily or taking recreational drugs can have irreversible effects that can be too late to 'repair' at birth. Heavy drug abuse during pregnancy can lead to the baby being born addicted to drugs, and ultimately experiencing withdrawal symptoms from birth. Different drugs will have different effects – the stronger the drug, the worse it will be. Consider the use of marijuana: when a pregnant woman gets 'high', the baby is also under the influence of the drug – at this crucial time when the developing foetus is trying to create healthy neural networks. As identified in a range of research studies (Lambe et al. 2006; Bada et al. 2005; Langley et al. 2005; Fried and Watkinson 2001; Levitt 1998; Chasnoff et al. 1987), other effects of substance abuse on the foetal development include:

- early delivery.
- low birth weight.
- miscarriage.
- hindered development.
- Foetal Alcohol Syndrome (FAS). This refers to a range of mental and physical defects that develop in a foetus as a result of high levels of alcohol consumption during pregnancy.
- defects of the face and body.
- intellectual disability.
- heart problems.
- death.

Pause for thought

1 A young woman who is expecting her first child asks you for dietary advice. What would you advise to help her promote her baby's brain development in utero?
2 Discuss two practical ways in which you can harness evidence from the first 1001 days campaign to inform your work with young children and their families.
3a Discuss one way in which epigenetics can inform your practice.
3b How far do you think your practice could be improved as a result?
4 In your own words, how would you describe the link between neuroscience and the nature versus nurture debate?

Understanding brain development as key to quality in early childhood education and care

Excitingly, the way in which we now recognize and view quality is being informed by our understanding of early brain development and neuroscience. This relatively new emphasis on the importance of early brain development (from conception) also means an acknowledgement of experience-expectant brain development. This at last places at the forefront the young brain as an active and enquiring structure which requires consistent nurturing for optimum growth and development. Concepts such as *experience-expectant learning, sensitive periods, nurturing relationships – how neuroscience is used to inform attachment aware practice* and *sustained shared thinking* all form part of the discourse regarding quality early childhood education and care.

Each of the four concepts concerning early childhood development and neuroscience will now be explored in turn. While reading, reflect on your practice with children aged up to three years. This includes the physical environment, the emotional climate, resources and quality of adult input during interactions with infants.

Experience-expectant learning

The human brain is primed to learn. That is, it *expects* stimulation from external sources such as sounds, visual images, tastes, touch and smells in order to facilitate the development of its visual, auditory, tactile and olfactory systems. As a result of this stimulation, the brain 'rewires' – with new neural pathways being established in response to the stimulation. Repetition of learning in turn leads to the laying down of pathways. This type of learning is known as experience-expectant learning. Consider the well-known experiment conducted by Hubel and Wiesel (1962). Their suturing of kittens' eyes during the first three months of life demonstrated the irreversible effects of cutting off external stimulation that is necessary for neural growth in the visual cortex. During adulthood when the kittens' eyes were opened, they found a reduced number of neurons reacting in the sutured eyes, as the neural circuitry became irreversibly altered during this critical period of experience-expectant learning. Although such findings derive from animal studies, they are also applicable to sensitive periods in human development. One common example used to explain sensitive periods for experience-expectant learning in neurobiology and human development is language acquisition. Infants (between six and twelve months old) who enjoy exposure to language(s) are better prepared to acquire the necessary skills to start speaking the language(s) spoken in their environment. Their ability to listen to the sounds, process meanings and distinguish between words all results in neural representations (mental images) being formed (White et al. 2013; Kuhl 2010; Thomas and Knowland 2009) and as a result, language learning taking place. This experience-expectant learning paves the way for learning increasingly complex rules pertaining to language such as syntax and grammar. White et al. (2013: 3) identify the effects of early exposure on the young brain when it comes to language development:

> During this time, which some view as the sensitive period for phonetic learning, exposure to the language(s) used in their environment is thought to guide infants' formation of language-specific phonetic representations that serve optimal processing of their native language(s). This exposure strengthens the neural representations for speech sounds in infants' native language(s).

The ideal time for experience-expectant learning to take place is between birth and approximately five years, with different skills having different time frames. Parents and practitioners alike can support early brain development by stimulating each of the five senses through during the daily routine. Giving eye contact during care routines not only helps to build trust but also facilitates young babies' developing eyesight as they learn to 'track' and focus with both eyes simultaneously. Given that we now understand that both genetic factors and experience-expectant factors play a role in developing and strengthening brain architecture, we should make every effort to optimize every child's learning experiences. This means taking into account that each child has different temperaments and will therefore respond very differently to the learning opportunities provided: how this is managed by practitioners can consequently shape educational outcomes for children. Although curricula frameworks vary internationally, most now place emphasis on the birth to three period, in recognition of this unique phase of human development. Such examples include the Early Years Foundation Stage (EYFS), the Reggio Emilia approach, the Montessori method, the Te Whariki Early Childhood Curriculum and the Highscope Pre-school Curriculum.

Pause for thought

1 Consider the concept of experience-expectant learning. Given that it generally occurs in the early years of childhood, why might it be more difficult to teach specific skills such as native language in late childhood?

2 Reflect on partnership with parents in your setting. How might
    practitioners support parents' understanding of:
    a) experience-expectant learning?
    b) their role in facilitating their young child's development on a day-
       to-day basis?
3 Think about the curriculum provided in your setting;
    a) Is there any reference to early brain development in the document?
    b) In which ways could this further improve quality of provision in
       your setting?

Top tips for practice!

- Reflect on the emotional climate in your setting. Are there any
  features which could be adapted to minimize stress in babies and
  young children?

- Consider the physical environment. How might this be modified to
  embrace a more brain-based approach to teaching and learning?

- Make those early experiences count for all babies and children with
  whom you work!

- Follow the children's interests – really take a close look at what
  babies and children are engrossed in as they play independently, and
  use this information to plan from.

Concluding thoughts

A recurrent message in this chapter is that nature and nurture should
not be pitted against each other but instead be viewed in terms of their
interrelationship. When we consider these two critical aspects of human
development as mutually dependent, we become more amenable to reflect on
our role in helping to create a positive start for all young children entrusted in
our care and especially those who have not benefited from a good foundation.

Epigenetics shows us that our genes are susceptible to external influences: this can mean good or bad news because as we know, these influences
can be positive or negative. Factors such as prenatal nutrition, the presence
of prenatal depression or chronic stress and maternal lifestyle choices can
each potentially alter DNA make-up, leaving 'scars' on DNA in the form of
epigenetic markers. Having low birth weight, susceptibility to depression,
anxiety and continually raised cortisol levels are just a few examples of

such epigenetic markers. Armed with recent insights from neuroscience about how the brain learns and those factors which *you* can influence might encourage a more brain-based approach to early childhood education and care which can pay dividends for children and their families in the long term.

Further reading

Dowling, J. E. (2004). *The Great Brain Debate: Nature or Nurture?* Princeton and Oxford: Princeton University Press.

Dowling's book comprises three parts – parts one and two cover the developing brain and the adult brain, while part three provides an examination of the aging brain. The in-depth discussions concerning brain plasticity, the role of nature and nurture as well as learning languages are each grounded in the latest brain research, and are easy to follow. This is a good book to dip in and out of, depending on your area of interest or need.

Meadows, S. (2016). *The Science Inside the Child*. Oxford: Routledge.

This book is an engaging and accessible introduction to child development from different scientific perspectives. Chapter 4 (genetics and epigenesis) and Chapter 5 (neuroscience) are particularly useful for those wishing to learn more about the influences of nature and nurture when considering child development.

Ridley, M. (2011). *Nature via Nurture*. London: Fourth Estate.

Ridley guides the reader through the complexities and contradictions surrounding nature and nurture, with links to Konrad Lorenz, Jean Piaget and Sigmund Freud along the way. This book is another good resource to dip in and out of, depending on your area of interest.

Chapter 6
Conclusions and Recommendations: Where Do We Go From Here?

This book has discussed a range of issues relating to early childhood education and care and neuroscience. As well as defining what neuroscience is, consideration has been given to some of the common myths that continue to permeate this discipline. Discussions were included to encourage reflection on issues such as the implications of these myths on early childhood education and care, and how practitioners can challenge these in their work with young children and their families. Stephen Porges' Polyvagal Theory, the limbic system and the effects of the stress hormone, cortisol, in young children's emotional lives were also explored in order to build readers' understanding of how these can impact on the development of self-identity and emotional well-being. Epigenetics was explored in relation to the complex issue of nature and nurture, with due regard given to the first 1001 days of life – a subject that will be of particular significance to those working in preventative roles as part of their profession. The ultimate aim of this chapter is to instigate a change in thinking regarding how early childhood practitioners might be able to embrace revised ways of working that are informed by neuroscience.

What to expect from this chapter

In line with the issues explored throughout this book, three related professional challenges (or topics) will be discussed in this concluding chapter. These challenges are:

1 How do we nurture children with poor socio-emotional development?
2 How do we support mothers with postnatal depression?
3 What can neuroscience offer to our thinking about the need for positive educative relationships as a key element in enabling learning environments for young children?

As part of each challenge, the following sub-sections will feature, in order to give structure to the concluding thoughts presented:

- A description of the challenge facing professionals.
- Aspects of neuroscience that the practitioner might think about or act in their practice in relation to this challenge.
- Questions to help you reflect on your practice in the 'Pause for thought boxes'.

It is fair to say that, at present, early childhood education and care is still some way off being able to really embrace information from neuroscience in daily practice. This is due to factors such as lack of consensus from neuroscience, insufficient collaboration across disciplines and inconsistent application of neuroscience-informed practice across early childhood practice. However, it is hoped that this book can make a contribution to embedding this knowledge more firmly within the grasp of early childhood practitioners so that their practice can benefit from the important insights neuroscience offers. The first challenge presented explores how practitioners can nurture children with poor socio-emotional development. Issues discussed include the importance of healthy attachments in early life and the legacy that poor attachments can leave on an infant's brain architecture and socio-emotional development. Difficulties with interdisciplinary collaboration are identified, with a case study that highlights some of these barriers alongside proposed solutions.

When reading, consider your current knowledge concerning children with poor socio-emotional development and how this could be advanced with evidence provided by neuroscience.

How do we nurture children with poor socio-emotional development?

A description of the challenge facing professionals.

Understanding the impact of poor socio-emotional development on learning

Infants enter the world utterly dependent on primary caregivers to help them regulate their feelings and emotions. Their only understanding of feelings is through experience – and some of these experiences can be visceral and overwhelming. The principal way in which infants communicate need, distress, hunger or tiredness is by crying. It takes time, patience and consistency of care for infants to develop an understanding of why they feel certain emotions and how to overcome difficult emotions. Only then can they begin to process and express difficult feelings in ways that are healthy and not detrimental. Cohen et al. (2005: 2) offer a comprehensive definition of early socio-emotional development:

> Social-emotional development includes the child's experience, expression, and management of emotions and the ability to establish positive and rewarding relationships with others. It encompasses both intra- and interpersonal processes.

These intra- and impersonal skills essentially involve the ability to understand one's own emotional state – and understanding other people's moods and feelings. They also involve being able to express feelings and show empathy. Healthy socio-emotional development does not occur in isolation, nor can it flourish where a child does not have access to stable and affirming relationships with adults, who can guide them through this fundamentally important aspect of their life. Children with poor socio-emotional development may experience difficulty in forming friendships, be withdrawn and experience difficulty in managing feelings of frustration and 'hit out'. They may also experience separation anxiety from their main carer even in familiar environments. Of course, most children will display a number of these behaviours on occasion, but when the behaviours persist, children need swift access to support in order to minimize the impact on their self-esteem. Available support networks include paediatricians, nursery teachers, early years practitioners, child psychologists and play therapists. What is lacking is the perspective from neuroscience, which can show the interrelated nature of the brain regions responsible for socio-emotional development and learning. This is identified by Bell and Wolfe (2004: 366):

> Recent cognitive neuroscience findings suggest that the neural mechanisms underlying emotion regulation may be the same as those underlying cognitive processes.

So, what does this mean? Consider a child at nursery who is generally well adjusted and sociable. Their limbic system (which includes the amygdalae and hippocampus) does not unduly become activated into flight or fight mode as the child is not stressed: their cortisol levels are stable. This means that key brain regions used in learning, like the pre-frontal cortex (responsible for problem-solving and planning), are not compromised by the surge of cortisol and adrenaline. These hormones are released when a child feels stressed, anxious or threatened, reducing the efficiency of the pre-frontal cortex. However, simply knowing which brain regions are responsible for socio-emotional development learning is not enough – on its own this is of limited use to practitioners as it does not explain *how* the brain regions function. This challenge is discussed by Mason (2009: 549):

Studies that are only technology-driven are unnecessary; instead, theory-driven studies are needed. It is, therefore, important to understand the function of the various brain regions involved in cognitive processing, especially in cognitive activities performed by children with different cognitive and motivational characteristics.

Making this knowledge meaningful to the children you work with will be the determining factor. This will require practitioners to be trained specifically in children's socio-emotional development from a neuroscience perspective so that they can put the evidence from theory-driven studies into their practice. This may include rethinking and re-evaluating outcomes for children which are informed by neuroscience. Below is a case study provided by the Chief Executive of a membership organization for early childhood trainers and consultants in the United Kingdom. Proposals to try to overcome the challenge of uniting neuroscience and early childhood education and care are also included. This case study gives prominence to two further challenges in uniting neuroscience and early childhood education and care:

- the urgent need for early childhood practitioners and neuroscientists to collaborate in trying to make evidence from neuroscience translatable and applicable in supporting children who have poor socio-emotional development.
- developing a national standard training course which includes a neuroscience module to equip practitioners in supporting children who have poor socio-emotional development.

Case study

Neuroscience is highly relevant to education. However, there is still a gulf between current neuroscience science and direct applications. The marriage between the science and education is in its very early days but with so much misinformation it is essential to take action now and create a clear pathway for neuroscience to become an essential aspect of teacher training and parent education. To understand the infant and the young child, a sound training in child development and observation is key to the early years teacher's success in supporting the highest possible outcome for children's early learning and development. Nurturing development socially, emotionally and physically as well as young children's individual progress is an absolute requirement to support and enable children's future life chances. This is especially true for disadvantaged children, children with special needs and/or disabilities, and for children who have faced trauma and abuse in their early years or even in the womb. The emergence of educational neuroscience has been born out of the need for a new discipline that makes scientific research practically applicable in an educational context. Neuroscience research has informed government education policy for children under three in many countries including the USA and the UK. We believe that neuroscience is an essential ingredient of our training courses and we are developing a national standard training course for our members – nationwide and internationally – to include a neuroscience module. Some of the topics we aim to include on the module are neuroscience informed attachment, epigenetics, maternal well-being and foetal brain development and the practical applications of neuroscience in early years provision.

Proposals such as those outlined in the case study may be deemed ambitious and will clearly require meticulous cross-disciplinary planning, with its delivery being monitored and assessed for effectiveness. That said, the approach is much needed in instigating a change in the training, thinking and practice of early childhood practitioners, which can consist of findings from neuroscience.

Children who experience difficulty in self-regulating their emotions and the impact of this on their all-round development and learning

Some children enter nursery able to regulate their own emotions. This might look like a child attempting to resolve conflict with a friend, or being able to tell a friend that their behaviour has hurt their feelings. However, many children begin nursery lacking these vital life skills and consequently find life at nursery very challenging due to difficulty in expressing their feelings and exercising self-control when they need to most (Bradley et al. 2009). This challenge is identified by Boyd et al. (2005: 2):

> Children enter nursery unable to learn because they cannot pay attention, remember information on purpose, or function socially in a school environment. The result is growing numbers of children who cannot get along with each other, follow directions or delay gratification.

Research shows that the gap in socio-emotional skills between children who live in poorer households and those who live in more financially secure households has doubled within a generation (Early Intervention Foundation 2015; Conti et al. 2011; Feinstein 2000). Some evidence points to children from disadvantaged backgrounds being more likely to begin nursery or reception with weaker self-regulation skills than their peers. Possible long-term effects of this can include diminished cognitive and socio-emotional ability and socio-economic success in adulthood. Research carried out by the Early Intervention Foundation (2015: 10–11) concluded that:

> [p]oorer children exhibit more conduct and emotional problems and that this gap appears very early in childhood. By age three poorer children

display worse conduct than their wealthier peers, and these differences persist throughout pre-adolescence.

Young children can therefore benefit from practitioners who are specifically trained in understanding socio-emotional development and how to instil qualities such as self-control, delaying gratification and decreasing impulsive behaviour. Support from attuned adults can help to ensure that the close interrelationship between a young child's ability to self-regulate and their cognitive abilities does not become compromised by factors beyond the child's control. Ultimately, strategies that have a preventative role can have a lasting positive impact on later learning at school, and on wider, long-term outcomes concerning behaviour, persistence and educational attainment.

Aspects of neuroscience that the practitioner might think about or act in their practice in relation to this challenge
In this section, references will be made to neuroscience-informed curricula frameworks and strategies that are recognized as effective in improving outcomes for young children. These will not be explained in great depth but instead serve as starting points for further research depending on your area of professional interest. Examples of good practice include curricula frameworks like the *Tools of the Mind Curriculum*. This is predominantly used in the US and its core aim is to develop self-regulation alongside skills in literacy, mathematics for children from low-income families and children with English as an additional language. The curriculum is based on the work of Vygotsky (1978), with an emphasis on play planning (done by the children with support from the teacher), learning through activities that are self-correcting and require self-regulation. This attachment and self-regulation approach to the curriculum has shown improvements in the quality of play, self-regulation and educational attainment (Barnett et al. 2008; Bodrova and Leong 2007; Diamond et al. 2007). The focused work on improving self-regulation within these frameworks is necessary as overstimulation can lead to problems for a young child, especially in busy settings where children can get overexcited by all that is on offer. Providing more focused time for learning, with an emphasis on fostering self-regulation, can result in calmer children who will find it easier to pay attention, plan, self-regulate and reflect.

The *Synapse School, neuroscience-informed nurture groups* and *Attachment Aware Schools* are also growing in popularity in some nurseries and schools across the United Kingdom and America. The primary purpose of these groups is to work with specific children from impoverished emotional backgrounds to strengthen their emotional, social and cognitive development. Each draws on elements of attachment-led relationships, emotional intelligence and brain-based and project-based instruction which are mainly constructivist in their nature. Practitioners draw on Bowlby's attachment theory and neuroscience to inform their work. Bennathan (2009: 1) explains the thinking behind neuroscience-informed nurture groups:

> Recent developments in neuroscience suggest that early infant experience, especially stressful and low nurturing environments, can have a negative impact on brain development and ... early intervention which helps to redress this via nurture, can have positive neurological effects.

Importantly, research has shown that the groups are effective in improving socio-emotional skills and cognition, alongside supporting parents to become partners in their child's education – which are critical in helping to help break the cycle of a poor emotional start in life and later youth offending (Commodari 2013; Bennathan 2009; Geddes 2006; Gottman et al. 1996). Similarly, Attachment Aware Schools develop children's emotional and socio-emotional well-being and improve academic outcomes through neuroscience and attachment-informed practices including emotion coaching, the support of family play inclusion workers and the encouragement of relational closeness with practitioners rather than distance at times of difficulty with the children. This last point is highly significant for a child who cannot handle their emotions and consequently has 'time out' or some other form of behavioural correction. Relational closeness can help an overwhelmed child to calm down and avoid feelings of anger and guilt. A significant achievement in the wider dissemination of Attachment Aware Schools is its recognition in the government's statutory guidance on promoting the well-being of looked-after children (2015).

The current Early Years Foundation Stage (EYFS) does have Personal, Social and Emotional Development (PSED) at its core, but this emphasis is lost when the child enters primary and secondary education, where the focus is on academic outcomes. With so many children requiring extra support emotionally, embracing new perspectives of learning from neuroscience-informed attachment could make a lasting difference to well-being and outcomes both in the short and long term. This is recognized by Siegel (2012: 311), who explains the role of supportive attachments in optimizing brain development:

> The attunement of emotional states is essential for the developing brain to acquire the capacity to organize itself more autonomously as the child matures.

Interdisciplinary collaboration and the dissemination of knowledge from neuroscience

The continuing gulf between neuroscience and early childhood practice is gradually being eroded by some knowledgeable and dedicated early childhood practitioners, lecturers and researchers. This is happening alongside increasing evidence from numerous cross-disciplinary studies such as health, psychology, economics and neuroscience (Almond and Currie 2010; Kolata 2007; Cunha et al. 2006). These confirm that early and continual investment in children's socio-emotional development is essential for optimal brain development and for our societies. They lay the foundations for our capacity to achieve and to function well despite social or even biological obstacles throughout one's life course. Results include neuroscience-based teaching on some early years programmes of study and practice. Such efforts are serving as catalysts for change in the theoretical frameworks embraced among early childhood practitioners. Psychoanalyst and clinical psychologist Professor Peter Fonagy discusses one persistent challenge in applying neuroscience to practice with regard to interventions for children's emotional and psychological disorders:

> I think that neuroscience is part of the answer and it is going contribute to solutions. The ultimate problem is we are trying to do too much which creates too many complications and which makes the interventions less effective. So by understanding the brain mechanisms more, we can become more precise in our interventions. (Fonagy 2015, personal written communication)

We are reaching a point where we can now incorporate discoveries from neuroscientific sources that are shedding more light on an infant's developing brain architecture and the impact of early care and learning experiences on this. Research carried out by Goswami (2015), Save the Children (2015), Karmiloff-Smith and Karmiloff (2015), Trevarthen and Delafield-Butt (2013b), Meltzoff et al. (2009), Hackman et al. (2009), Brotherson (2009), Immordino-Yang and Damasio (2007) and Fonagy (2001) have much to offer to build on existing knowledge in this important area. Making these distinguished names in neuroscience more accessible across the early childhood sector can make a huge difference to how practitioners comprehend and consequently choose to guide our youngest citizens in their relationships and their developing self-identity. Increased accessibility to these theorists' ideas will thus require their wider dissemination across early childhood education and care qualifications and practice.

Pause for thought

1a How do you help young children to develop their emerging awareness of self and others?

1b How might your approach be developed in light of your

understanding about the interrelated nature of emotional and cognitive processes?

2   What is the link between early brain development and socio-emotional development?

3a How might practitioners benefit from understanding the interrelationship between self-regulation and cognitive outcomes for young children?

3b Discuss three practical ways in which this knowledge can inform your daily work with young children.

How do we support mothers with postnatal depression (PND)?

A description of the challenge facing professionals

Using neuroscience to understand the effects of maternal stress and depression on foetal and infant brain development

Postnatal depression (also known as 'baby blues') affects approximately one in every ten women in the United Kingdom within a year of giving birth. PND includes a range of symptoms that can vary in severity. Mothers with PND typically experience the following:

- Feeling little or no joy when spending time with their baby.
- Difficulty in forming a bond with their baby.
- Persistent feelings of hopelessness and sadness.
- Feeling tearful a lot of the time.
- Feeling anxious.
- Constantly feeling tired and lacking in energy.
- Experiencing invasive thoughts like harming their baby.
- Difficulty in 'thinking clearly' or making decisions.

Trevarthen et al. (2003: 10) remind us of the possible consequences for an infant's emotional and social well-being if their attempts at engaging in a relationship are not responded to by their mother. Of course, a one-off or a 'bad day' may not cause distress in the long term. It is the ongoing lack of affectionate relationships that can leave an infant chronically withdrawn.

An intimate and affection relationship is one in which emotions are shared freely and strongly. If there is lack of intimacy, if the mother is too tired and depressed or worried to care for her infant, this causes the infant to protest and then withdraw.

The Royal College of Midwives (2012: 3) makes reference to neuroscience and early brain development concerning the role of healthy attachments in building healthy brains:

Evidence about early brain development has highlighted the importance of building a bond with the unborn baby.

From the moment of conception, the mother's feelings towards her unborn child can have long-term effects on their relationship when he or she is born. Simple gestures like talking to her 'bump', stroking it, playing soothing music and trying to avoid chronic stress can all benefit both mother and baby. In one study, Benoit et al. (1995: 310) found that expectant mothers who had positive 'representations' (thoughts) of their unborn baby enjoyed more a secure attachment:

The richness of ante-natal maternal representations was significantly linked with the security of the infant's attachment at one year of age.

Of course, it is not 'make or break' if the mother does not feel positive during pregnancy – such a life-transforming experience can cause intense anxiety, pressure and uncertainty. This does not preclude feeling love towards her unborn baby. It is when negative feelings persist and inhibit a healthy bond forming that support may prove necessary. A mother who is emotionally available and able to meet her infant's needs for companionship can help them to build positive ideas about relationships and emotions. Where a mother is not emotionally available, perhaps as a result of being too depressed, she may experience difficulty in managing her own internal emotional life and in responding appropriately to her baby's emotions. Increased understanding of mental health conditions with regard to brain development can gradually act as a catalyst for psychiatrists, neuroscientists, researchers and practitioners to bridge the current gap in practice: a challenge identified by Professor Peter Fonagy:

In some studies, researchers go back much further and study birth and even prenatal indicators of mental health conditions. Social and psychological impact will always be experienced at the level of the brain. I think some of our failure in psychiatry comes from the inadequate use of neuroscience. (Fonagy 2015, personal written communication)

Given that practitioners play a supportive role for families, being knowledgeable about PND could serve as a welcome support mechanism where insecure or avoidant attachments exist between a mother and her child. The attachment bond is a key factor in the way an infant's brain organizes itself and, in turn, influences their emotional, social, intellectual and physical development (Tronick 1989). The National Scientific Council on the Developing Child's Working Paper 2 (2010b: 4) asserts that:

> Young children who grow up in homes that are troubled by parental mental-health problems, substance abuse or family violence face significant threats to their own emotional development. The experience of chronic, extreme, and/or uncontrollable maltreatment has been documented as producing measurable changes in the immature brain.

Aspects of neuroscience which might help the practitioner think about or act in their practice in relation to this challenge

Understanding the child's intentions and feelings for a creative life in affectionate companionship

An infant's impact on their world starts from life in utero. Four-dimensional scans (which show moving 3-D images of the foetus – time being the fourth dimension) reveal that from the second trimester, the foetus demonstrates *conscious* movement which is shown through its *purposeful actions*. This includes moving their legs or arms when they hear their mother's voice or to touch an object: in this case, the wall of their mother's uterus. This acting with knowing (i.e. reaching out and moving to touch and make sense of their surroundings) is known as 'intentional' and 'goal-directed' (Delafield-Butt and Gangopadhyay 2013; von Hofsten 2007; Trevarthen 1978). As the foetus continues to develop, so does its actions, which become more sophisticated. This can include making facial expressions (smiling or frowning) in response to soothing or startling noises or pushing its elbows or legs to move position. These movements are often affectionately felt, seen and responded to by pregnant mothers as 'I think this one will be a footballer!' or 'Someone is feeling restless!' Such affectionate utterances to her unborn child pave the way for the attachment, communication and relationship to begin its lifelong journey.

How do these foetal intentions and feelings in movement manifest after birth? Well, remarkably, all this purposeful action experienced in utero leads to the infant's fascinating ability and readiness to be physically close to and interact with her mother. It is these intentional movements that become instrumental to an infant in initiating relationships and being an active partner in conversations with parents and other close adults. Although all of a baby's needs must be met by a caring adult, this does not mean that they are completely passive 'blank slates' socially and emotionally. They are incredibly curious and eager for conversation and companionship (Trevarthen et al. 2003; Trevarthen and Aitken 2001). This is recognized by Makin and Whitehead (2004: 14):

> Babies are born already prepared to find other people interesting and worth communicating with from the start.

This means that all the singing, action-rhymes, games (like peek-a-boo), affectionate gazing, sharing of stories and early conversations and mimicking each other's facial expressions encourage affectionate companionship between the infant and their parents, or other adults important to them.

What relevance does understanding foetal intentions and feelings in movement have for practitioners?
Professor Colwyn Trevarthen's vast body of compelling research spans over four decades and is renowned globally. His work primarily concerns infant brain development, communication and emotional health, and how babies use rhythm and expressions in their movements to help communication with significant others. His work (and collaborative work) can be practically useful to parents, teachers and therapists, by stimulating a change in thinking. This includes how we observe, interpret and respond to babies' attempts at communicating and when giving them care and companionship. When practitioners tune into an infant's need and desire to be active in their social endeavours, they can follow their lead and enjoy a happy and healthy relationship with them. Trevarthen et al. (2003: 132) identify the benefits of quality practice on a young child's emotional development – which is all the more important where there is a history of maternal mental health problems:

> An ideal practitioner will give consistent and sustained responses to each child's individual personality and history of experiences. Maternal care, with its natural benefits to child and mother, can be substituted only by persons who know how to respond with the consistent intimacy, care, affection and enjoyment that a happy mother can give to a loved child.

The more practitioners understand early development, the better they can support parents in making a connection with their babies. This is not just a case of anxiety about the infant's survival and need for love and care, but more about how they can enjoy the companionship of their baby, who enters the world so willing to share with caring and affectionate others. Trevarthen and Delafield-Butt (2013b: 1) conclude that:

> Understanding the root of narrative in embodied meaning-making is important for practical work in therapy and education, and for advancing philosophy and neuroscience.

One child play therapist details the effects of PND and related maternal mental health problems through the case study that follows, which discusses an infant who, as a result of her mother's mental health problems, experienced pre-verbal neglect and trauma from just months old. This therapist harnesses neuroscience-informed play therapy, which has shown the positive effects of symbolic and fantasy play on the young brain to form new neural pathways in response to the play experiences (Modell 2003; Levin 1997). The case study outlines how he was able to read and interpret the child's internal state due to his deep understanding of brain and language development, and how pre-verbal trauma manifests through non-directive play therapy.

Case study

As a play therapist who specializes in treating the youngest victims of abuse, neglect and domestic violence, I have witnessed first-hand the resilience of many of my clients and their capacity for processing trauma through play. However, one case in particular stands out, as it not only serves as a testament to the potential for healing through non-directive play therapy, but also an example of the treatment of pre-verbal trauma through play therapy.

Melissa was four years, two months of age at the time of referral. She was in a foster care placement, having been removed from her biological mother six months earlier. This was her second removal, the first taking place when she was eighteen months old. During the initial consultation, the foster parent reported Melissa was exhibiting severe tantrums at home and in her preschool classroom, triggered by circumstances that compromised her low frustration tolerance and high levels of anxiety. Coupled with her limited capacity for self-regulation, Melissa had a very hard time modulating her emotions. She was also prone to physical aggression, often hitting her peers in response to the slightest conflict. Melissa was in the process of being evaluated by her school district's Child Study Team to assess the presence of a learning disability and her need for special education.

The case history noted Melissa's previous exposure to neglect. At eighteen months of age, she suffered from diaper rash that was so severe it required hospitalization. However, while she was treated for her physical ailment, she never received any type of developmentally appropriate therapeutic support for her emotional wounds, which in my estimation were likely influencing her challenging behaviours two and a half years later.

Melissa participated in individual play therapy, one session per week for 45 minutes. My primary treatment approach is non-directive play therapy. My play room contains many of the materials consistent with that of the typical play room. However, I have integrated other items into my room that deviate from the standard child-centred list, including a sand tray and miniatures, foam swords, and an array of Playmobil vehicles and people. For children who have been removed from their homes or moved, I also have three different types of dollhouses in the room to facilitate relocation play. Many times I have witnessed symbolic play scenarios where a child removed the contents from one house, placed them in a vehicle, and 'drove' them to another house where they proceeded to furnish the new home. Another essential addition to my playroom are pretend metal handcuffs. I have found that for a child who has been neglected or exposed to domestic violence, the opportunity to handcuff the therapist satisfies their need for control and is highly corrective.

In initial play sessions, Melissa explored the play room tentatively. She was equally tentative in her engagement of me. I never use a chair in my playroom. In order to empower the child to the fullest extent possible, I sit on the floor and try to look as 'small' as possible. While this approach often has a tremendous influence on the child's trust of the therapist, Melissa needed a higher level of control and thus handcuffed me at various points during the first three sessions.

As her trust of me increased, so did the developmental appropriateness of her play. Melissa engaged in pretend play where she was the mother and I was her child. She dutifully prepared pretend meals and she fed me. The symbolism for this type of play, particularly when it involves a child who has experienced prolonged maternal neglect, is unmistakable. However, after the seventh play therapy session, Melissa initiated a play sequence that was profound. At about the fifth week, Melissa, at some point in the session, knelt before the dollhouse. She took the small infant, held it in her hands briefly, and placed the baby face down. That was the extent of her engagement. She then proceeded to the next play activity.

For the next four sessions, Melissa repeated the act of selecting the dollhouse baby and placing her face down in the crib. She also began a similar sequence with the large baby doll. She would choose the doll, hold it briefly, then place it face down in its small fabric carrier. She never verbalized during either of these play expressions. At this point, I reflected on the possible meaning of Melissa's play, particularly her silent, repeated placement of the baby's face down. I called the caseworker and asked for a more detailed medical report regarding Melissa's hospitalization several years earlier. Within hours, I had the answer: Melissa's diaper rash had been so painful she was unable to sit on her bottom or sleep on her back. For the tenth session, there were several items that would be waiting for Melissa to utilize if she chose: a bowl of water, a miniature bar of soap, and a wash cloth. Upon entering for the tenth session, Melissa integrated the bath items into her play as I had hoped she would. She promptly gathered the baby in her arms and sat before the bowl. Resting the baby in her lap, Melissa took the washcloth in her hand and dropped it in the water. She alternated between rubbing the soap on the baby's bottom and wiping it with the wet washcloth. Melissa was washing the baby's bottom, an act of nurturance she had not experienced from her own mother. Melissa repeated the washing play for three consecutive sessions. At one point during session fourteen, I witnessed two play expressions that today I consider among the most powerful I've ever witnessed in my years as a practising play therapist. Upon choosing the baby, which had now become 'her' baby in a parent–child role play, Melissa washed her baby's bottom, kissed her on the forehead and placed her on her back side in the carrier. She then approached the dollhouse, selected the infant doll, and placed her in the crib on her back side.

Throughout the course of Melissa's play therapy, I held routine telephone consultations with her caseworker, foster parent and preschool teacher. By all accounts, Melissa's behaviour had significantly improved. In fact, her tantrums and physical aggression had diminished to near zero levels, she was interacting appropriately with her peers and her ability to tolerate frustration appeared developmentally appropriate in and out of the playroom. Melissa was no longer considered a candidate for special education and the recommendation for medication to treat her behaviours was withdrawn. Melissa had become a happy, well-adjusted four-and-a-half-year-old girl who was comfortable in the nurturing, structured and consistent environment of her foster care placement and demonstrating an age-level ability to learn and socialize in her preschool classroom. While there were more therapeutic matters that needed to be addressed, including the eventual commencement of visitation between Melissa and her biological mother, which would be therapeutically supervised by Child Protective Services, Melissa was better prepared to cope with the potential stress and anxiety of the visits. Melissa's traumatic hospitalization occurred at a point in her development where her language was just beginning to emerge, as she was uttering one to two word phrases as per the case history. Research suggests that the capacity to retain meaningful internal representations of the salient elements of a traumatic experience may be present as early as the second half of the first year of life. Thus, this case serves as an example of the need to treat children who experience pre-verbal trauma with developmentally appropriate therapies. While it is very difficult to establish a correlation between Melissa's play therapy and the dramatic improvement in her adjustment and behaviour, this case is a compelling example of the potential efficacy of treating a child who experienced pre-verbal trauma through non-directive play therapy.

Although it is not outwardly clear, this case study has outlined how this play therapist helped to create new neural pathways in the child's brain. This was achieved over time by calming the excessive activity of the amygdalae, which, as a result, helped to stop feelings of anxiety and aggressive behaviour in the child. Ultimately, the play therapy sessions helped to establish new coping mechanisms that will prove invaluable throughout this child's life.

The role of PET scans in identifying excessive levels of enzymes causing PND

PET scans are being successfully used to identify increased levels of the enzyme monoamine oxidase A (MAO-A). High levels of MAO-A pose a mental health risk for mothers in the period just after childbirth, as it can cause PND and the associated feelings of sadness, anxiety and depression. This is because high levels of MAO-A breaks down the neurotransmitters serotonin, dopamine and norepinephrine. Ground-breaking research carried out by Sacher et al. (2010) has for the first time demonstrated this connection and it is hoped that these findings will lead to the development of critical

interventions which can prevent PND from occurring. Proposed interventions include:

- Creating dietary supplements that counter the breakdown of the neurotransmitters when MAO-A levels are high.
- Keeping levels of MAO-A levels balanced in the brain by either lowering levels or by increasing levels of serotonin, dopamine and norepinephrine.
- Developing treatments that are compatible with breastfeeding.

Attempts can be made to either lower elevated levels of MAO-A with selected drugs, or to increase the concentration of monoamine neurotransmitters that can lift mood. Given the need to develop treatments that are compatible with breastfeeding, the intake of dietary supplements of MAO-A precursors in the early post-natal period looks like a promising strategy during this important time for a mother and her baby.

Pause for thought

1 Discuss two ways in which you could use research from neuroscience in your practice concerning PND and its impact on babies' need for companionship.
2 How can evidence from neuroscience help you to maximize opportunities for children to access calming, sensory-rich play?
3a Consider your setting's learning outcomes for children aged three years and under.
3b Which outcomes do you place the most importance on?
3c Could these outcomes be adapted to place more emphasis on the emotional well-being of children from birth to three years old?
4 How might family-focused interventions help to improve the mental health of mothers with depression and the mental health and well-being of their children?

What can neuroscience offer to our thinking about the need for positive educative relationships as a key element in enabling learning environments for young children?

A description of the challenge facing professionals.

Using neuroscience to inform planning of enabling learning environments

The brain is essential for learning, that is, processing information from external stimuli through the five senses, and interpreting this information. Learning, then, is the result of this information processing and making meaning. It is therefore only natural that we give consideration to the conditions in which we expect the brain to learn and flourish. *The question is, do practitioners make the most of their setting's indoor and outdoor environments to nurture children's development and learning?* This means paying attention to the environments' physical, emotional, social, cultural and intellectual characteristics. The OECD (2007: 14) highlights the need to take into account how the brain works when planning learning environments.

> In the design of learning environments, new knowledge about how the brain works, reinforces holistic approaches that recognize the close interdependence of physical and intellectual well-being and the close interplay of the emotional and cognitive areas of development.

One head teacher of a local authority state funded nursery school and children's centre gives a clear account of how staff in the setting utilize knowledge from neuroscience and brain development in order to plan delivery of the EYFS.

Case study

Neuroscience presumably covers all we do with young children in early years. The best descriptions come from observing children's play and understanding key ways that neuroscience supports our practice.

- Attachment and attunement – we help children to settle into nursery and separate from a familiar caregiver (usually a parent) and make a bond with a keyperson. All our staff understand the need for appropriate attachments to help children emotionally develop – and develop emotional intelligence.
- We support children to self-regulate their behaviour. This means identifying what emotional state they are in and eventually returning to a state of well-being.
- Practitioners are able to select and use the right strategy to resolve a situation. For example, a six step approach to solving problems and resolving conflicts – using action words to help children connect with a feeling and what to do about it. This might be 'big' feelings – running to express anger or cuddles when they are feeling sad. Circle time activities are used to discuss feelings, using feelings cards and feelings games. We sing feelings songs like 'if you're happy and you know it' – but we change the words to reflect a range of emotions and what we can do when we feel like that.
- Practitioners recognize that they are a role model and they are able to calmly and

consistently help children self-regulate and restore a sense of well-being in difficult and emotional situations.

None of our staff have qualifications which covered neuroscience or early brain development. I think neuroscience should form part of the revised childcare qualifications because the more we know about brain development, the better support we can give to our babies and children. For now, our understanding comes from our ethos and curriculum provision which are based on the Reggio Emilia approach, forest school theory from the Scandinavian kindergartens and the Leuven scale of well-being and involvement. All three of these approaches have roots in neuroscience.

Embedding neuroscience and updated cognitive theories in early childhood curricula

The case study above highlighted that, generally, while practitioners do mention neuroscience as playing a key role in early learning, no formal training had been undertaken. One *positive* aspect associated with making such references is identified by the OECD (2007: 229):

> Teachers are not aware of the actual processes in the brain that constitute actual learning itself, and can at best only make some guesses about the mystery of brain mechanisms. It is therefore reassuring that brain and educational research validates such guesses.

Although the 'guessing' is perceived as reassuring here, it certainly is not enough on which to base neuroscience-informed early childhood practice, if it is to be reliable and of high quality.

It is not only practitioners who do not consistently make links to brain function in young children and the effects of the learning environment on this. They are not always clearly defined in early years curricula either – this can be viewed as another barrier to adopting revised ways of working that are strengthened by neuroscience. That said, curricula for early learning such as the EYFS (2015), Te Whariki, Montessori Method and Reggio Emilia approach each pay close attention to the learning environments. Although each has its distinct origin and rationale, their central tenets do converge. They all emphasize the role of the following:

- Resources are set up in ways that encourage independence.
- Indoor and outdoor physical spaces are created to allow for freedom of movement, exploration and choice.
- Social skills and interactions are actively promoted by staff.
- Young children are encouraged to use their own initiative when problem-solving.

- Intellectual development is nurtured through careful observation and planning of individual babies and children as they pursue their individual interests.

When considering the bigger picture, these learning frameworks are actually influenced by theories that are (commonly) adopted globally. Examples include Bruner's and Vygotsky's socio-constructivist theories of learning, which still frame pedagogical discourse today. Although their philosophies are extensively deliberated and adopted in primary, secondary and further education systems, they are also embedded in early childhood curricula. Their emphasis on the social and cultural climate of the learning environment and how this impacts on learning is acknowledged by Bruner. He believes that how we view the mind influences how we comprehend its interactions with the environment (Bruner 1996). The psychological processes involved in learning and the impact of the learning environment on these processes is widely recognized. Take the role of socialization in learning: children's brains are primed for socialization, and learn effectively in collaboration with others (UNESCO 2012; Adolphs 2009; Goleman 1996). Brain regions such as the mirror neuron system, prefrontal cortex, amygdala and hippocampus and are all deeply involved in the learning process, becoming activated in response to stimuli received. Imaging studies show that activation of the prefrontal cortex is related to activity in the amygdala – and that during times of anxiety and stress, prefrontal cortex function can be greatly diminished, resulting in reduced learning ability. There are social influences on the brain that have a direct impact on its ability to function optimally for learning. In the past two decades, infants have been increasingly recognized as seekers and providers of social interaction and communication. This requires a naturally rich and stimulating environment in which social interaction is prioritized. Accordingly, designing environments that are conducive to building healthy brains is imperative if we expect our youngest and – in some cases – most vulnerable children to learn. How we articulate this knowledge about the young brain and neuroscience within early childhood practice is, however, problematic. Many practitioners can talk about their understanding of the connection between children's emotional states and their impact on learning, but as the case study below shows, they do not make explicit links to the brain.

Aspects or theories of neuroscience which might help the practitioner think about or act in their practice in relation to this challenge
The case study below is provided by the head teacher of a state maintained nursery school and children's centre. A range of strategies are listed in the example of brain-based early childhood practice. The purpose of the case study is to make explicit some of the ways in which you can build on your existing practice.

Case study

Most quality early years in-service training now has some (often brief) references to neuroscience and brain development. We are a state maintained nursery school so we have five training days per year – like primary and secondary schools – in addition to any training staff undertake elsewhere. Sometimes we have external trainers come in to train all the staff on a chosen topic – and they will often refer to neuroscience and brain development during the session. Also neuroscience and brain development have been referred to by early years consultants when delivering training to us on the benefits of outdoor play including Mindstretchers and Forest School Training.

Our ethos and curriculum provision is mainly based on the Reggio Emilia approach – much of Loris Malaguzzi's work has links with neuroscience. The approach fosters children's ability to critically think and problem-solve using natural materials, loose parts, loose materials and the natural environment – rather than using coloured plastic toys which have limited uses and properties. In our setting, we are working towards having less and less coloured plastic and more open ended resources to deepen learning opportunities. We also have a resident artist on our staff team and a resident musician who also uses lots of theory rooted in neuroscience.

For our outdoor approach, we use lots of the Forest School theory from the Scandinavian kindergartens. In general we want to support children with attachment, attunement and developing a good sense of well-being, particularly as we keep being told that English children are often found to be the unhappiest in surveys. For this we use the (Belgian) Leuven scale of well-being and involvement. All three of these approaches have roots in neuroscience. Now seen as established good early years practice – we use the Leuven scales for measuring emotional well-being and involvement. We have half-termly meetings to discuss children with low well-being and involvement levels so that we can put in strategies for support. Learning is closely linked to emotional well-being. Children who have a low sense of self-esteem and well-being are not able to learn effectively. We need to support them to enable them to make appropriate progress in their learning.

The outdoor environment is increasingly viewed as a valuable aspect of the learning environment. Practitioners are thinking more creatively and deeply about *how* the outdoor environment can be used to promote young children's well-being and disposition to learning. The British Heart Foundation (2016: 1) highlights the connection between physical activity and early brain development:

> 91% of children aged 2–4 years do not meet the Chief Medical Officer's physical activity guidelines for their age group of three hours per day. We're concerned because these children are missing out on opportunities for health and development. It's proven that physical activity for young children

supports brain development, enhances bone and muscular development and benefits social and cognitive skill development and emotional wellbeing.

An example of good practice is outlined in the case study below.

Case study

A nature deficit environment affects emotional well-being. Many children nowadays do not have the same access to outdoor play opportunities. This might be due to excessive traffic or parents' awareness of stranger danger – and its prevalence in the media – or because parents work long hours and do not have time to take children outdoors. Lots of children live in urban areas with little access to nature. Children need nature for health, happiness and learning. They need nature-filled play areas. Even in our tiny urban outdoor play area we try to add as many opportunities for exploring nature as we can. We use environmental based learning techniques and 'forest school' activities to help children connect with nature. Through this type of learning they find out about:

- the natural environment and changing seasons;
- how to handle risks;
- how to challenge thinking;
- how to use their own initiative to solve problems;
- how to cooperate with others and learn boundaries of behaviour; both physical and social;
- how to grow in confidence, self-esteem and become self-motivated to learn.

These are all skills children need to develop to become effective learners.

Inside or outside, children can give vent to pent up feelings by using natural materials that cannot be damaged or lose their form if children pull and pummel them, like clay, play-dough, sand, water, soil and mud. Children can create and build or be destructive – they cannot fail when they use these materials. Underpinning all that children do in our setting are the three characteristics of effective learning as detailed in the Early Years Foundation Stage Framework:

- Playing and exploring – We observe how children use their hands (and other bits of their bodies) to play and explore.
- Active learning – We observe how children use their heart to want to learn.
- Creating and thinking critically – We observe how children are using their heads to really think.

Pause for thought

1a In which ways do you think your setting's learning environment affects young children's dispositions to learning?
1b In which ways can knowledge concerning brain development inform planning of your setting's outdoor environment?
2   How far could knowledge concerning brain development improve the quality of enabling learning environments in your setting?
3   What are the limitations in trying to apply knowledge about early brain development in your designing of enabling learning environments?

Concluding thoughts

One challenge which underpins this book is that of achieving interdisciplinary working in translating research from neuroscience into practice. The fields of psychology, early childhood education and care and neuroscience would each benefit from making steps towards working more cooperatively in early education and intervention (Howard-Jones 2014). Mason (2009: 549) recommends:

> The sooner educational psychologists invest their efforts in building a solid bridge, the more pedagogical practice and neuroscientific research will benefit from each other.

This collaboration can prove highly effective for improving outcomes for young children due to the traversing themes that they are able to join together over. This recommendation is supported by Koizumi (2004: 135), who introduces the term 'trans-disciplinary' when referring to the possibilities when neuroscience, education and other associated disciplines join together to share their expertise:

> Fusing neuroscience, education and other relevant disciplines, and creating a new trans-disciplinary field would connect work on learning across the intellectual walls dividing disciplines.

Achieving effective interdisciplinary working is therefore a necessity in supporting young children's development and learning, but it is often problematic to achieve. Issues commonly arise concerning collective vision and objectives, sharing a common purpose, conflicting work schedules, lack of understanding concerning roles and insufficient monitoring and assessment of interdisciplinary working. Such challenges are compounded by an insufficient systematic framework around which interdisciplinary working can be structured, monitored and evaluated. In relation to this, *translating* the information available into applicable interventions across early childhood provision (care, education and intervention) is also a challenge. For example, persistent questions concerning some types of brain scanning techniques such as Transcranial Magnetic Stimulation (TMS) mean that their use with young children is doubtful. Ethics, cost and long-term effectiveness in improving educational outcomes will take time to demonstrate, settle and refine. Howard-Jones (2014: 38) explains:

> Positive effects are now being reported for learning tasks relevant to education, but remaining questions regarding risk and ethics makes TMS classroom interventions unlikely in the near future.

This comment does not need to be interpreted negatively, but it does remind us that time and persistence are fundamental in the development of suitable research techniques for use with children. In addition, evaluation, continual refining and collaboration across neuroscientists and practitioners will need to be carefully coordinated in order to ensure that the application of neuroscience leads to theories that can be applied to early childhood practice. It will also take financial investment and determination until we can reach a point where practitioners, both nationally and internationally, are confidently using neuroscience-informed theories in their daily work of caring for and educating young children. In a recent heartening leader in the *New Scientist* (2016: 15), entitled 'If at first you don't succeed', we are told:

> Perhaps the most valuable weapon science has is its ability to be wrong – and with time and patience do better. Progress might be slow, and it might not happen in the way we like, but the eventual joy is worth any amount of torment.

Glossary of Terms

Attention deficit hyperactivity disorder
  (ADHD) is a lifelong condition that is often
  characterized by a group of behavioural
  characteristics including:
  ● Fidgeting.
  ● Restlessness.
  ● Being easily distracted.
  ● Having a short attention span.
  ● Being impulsive.
  There is no cure for ADHD but it can be
  managed when adults are aware of the
  condition and have an understanding of
  what interventions the child responds to
  best.

Autism is the generalized term given to the
  complex irregularities which occur in very
  early brain development. Autism tends to
  be characterized by difficulties in verbal
  and non-verbal communication, social
  interaction and repetitive behaviours.
  Some individuals may experience
  difficulty in maintaining attention and
  in motor coordination or may excel
  intellectually.

Axon is the long fibre that takes information
  away from the cell body of a neuron
  towards the axon terminals where
  connections, called synapses, form with
  neighbouring neurons.

Brain is the mass of nerves located in the
  skull. It controls everything we do:
  receiving, organizing and distributing
  information from the body. It also controls
  voluntary actions (walking, reading
  and talking) and involuntary reactions
  (breathing and digestion).

Brain Gym is a programme of physical
  exercises which claims to improve
  brain function through a set of exercises
  that are designed to encourage the two

hemispheres of the brain to work in
  synchronization. The programme has been
  criticized for its lack of scientific evidence
  and critical evaluation of classroom
  methods to support its claims.

Brainstem is situated beneath the limbic
  system and connected to the spinal cord.
  The brainstem is responsible for basic vital
  life functions such as heartbeat, breathing,
  blood pressure and swallowing.

Cerebellum (often referred to as the 'little
  brain') is made of two hemispheres. It
  has a highly folded surface area. The
  cerebellum is associated with regulation
  and coordination of balance, posture and
  movement.

Cerebral cortex is the layer of grey matter
  over the two hemispheres of the brain
  (often referred to as 'grey matter'). It is
  divided into four sections (lobes), these
  being the frontal lobe, parietal lobe,
  occipital lobe and temporal lobe. The
  cerebral cortex is densely packed with
  neurons and plays an integral role in
  consciousness, memory, intelligence and
  language.

Cerebrum (or cortex) is the largest part of
  the human brain and is associated with
  higher brain function such as thought
  and voluntary movement. The cerebrum
  is divided into four regions (lobes).
  These are the occipital, temporal, frontal
  and parietal lobes. Each lobe performs
  different tasks, but they must all work
  together.

Computerized tomography (CT scan) is a
  non-invasive diagnostic imaging procedure
  that uses a combination of X-rays and
  computer technology to produce images of
  the body or specific organs (for example,

the brain). During a CT scan, an X-ray beam moves in a circle around the organ. This allows many different views of it. The X-ray information is sent to a computer that interprets the X-ray data and displays it in a two-dimensional (2-D) form on a monitor.

Cortisol is a neurochemical that is produced in the adrenal glands. Generally, our cortisol levels are at their highest just after waking up in the morning, and the lowest levels are in the evening, before falling asleep. Cortisol is also released in response to fear or stress. Elevated cortisol levels can interfere with memory and learning, lower immune function and increase blood pressure and cholesterol.

Dendrites are the threadlike extensions on a neuron. They bring information (in the form of electrical messages) to the cell body of a neuron, in order for the cell to become active.

Depression is a mood disorder that can last for weeks or months. Symptoms include feeling sad, empty, hopeless, anxious, tearful, having no motivation or interest in things, finding it difficult to make decisions and feeling irritable and intolerant of others. Depression can be diagnosed as mild, moderate or severe. Other types of depression include post-natal depression, bipolar disorder and seasonal affective disorder (SAD).

Disequilibrium refers to an individual's inability to fit new information into their existing knowledge base; this puts the individual in a state of cognitive imbalance (disequilibrium).

DNA methylation refers to the epigenetic mechanism that can regulate gene expression without changing the underlying sequence unit of DNA. DNA methylation is critical to normal development.

Dopamine is a neurotransmitter that is essential for the normal functioning of the nervous system. It plays an important role in reinforcement and reward, executive functions, motor control, motivation and arousal.

Empirical research is based on information that is acquired by carrying out systematic observations, assessments and experiments, as opposed to relying on theory alone.

Epigenetics refers to external alterations to DNA that turn genes on (activates them) or off (silences them). These changes do not change the DNA sequence, but instead affect how cells 'read' genes.

Equilibrium occurs when an individual can transform incoming information so that it fits within their existing thinking; this means that they are in a state of cognitive balance (equilibrium).

Foetal alcohol spectrum disorder (FASD) is a condition caused by heavy drinking during pregnancy. FASD can cause poor growth in the womb. Children born with FASD tend to experience difficulty in functioning and coping with daily life. Difficulties are experienced due to the range of learning disorders that often result such as problems with thinking, speech, social skills and memory.

Functional magnetic resonance imaging (fMRI) is a non-invasive technique used to measure brain activity. It detects the changes in blood oxygenation and flow that occur in response to neural activity. Activation maps are produced from fMRI – these show which parts of the brain are involved in a particular mental process.

Gene expression is the process by which genetic instructions are used to create gene products such as ribonucleic acid (RNA) or required proteins. The main role of RNA is to act as a messenger, carrying instructions from DNA for controlling the production of proteins.

Gyri (bumps) and sulci (grooves) are what we can see on the brain's surface. Each of the two cerebral hemispheres is divided into four lobes by sulci and gyri. The 'folding' created by the gyri and sulci increase the amount of cerebral cortex that can fit in the skull.

'Ice pick' method (lobotomy) got its name after the doctor (Dr Walter Freeman) who pioneered this technique. The ice pick was refined and strengthened for this psychosurgical procedure in which the connections to the prefrontal cortex and underlying structures were severed or the frontal cortical tissue was destroyed. It was done by entering the top of the eye socket to access the brain and tapping lightly on the instrument with a hammer to break through the thin layer of bone and sever the fibres of the prefrontal cortex.

Interdisciplinary working refers to a coordinated group of professionals from several different fields who work together to achieve a common goal.

Internalizing behaviour involves adding attitudes, values and the opinions of others into one's own sense of self or identity. The term internalizing can also be used when referring to the fact that children and adults alike can also internalize problems, which means keeping their worries to themselves.

Intersubjectivity refers to the shared meaning that is created between an infant and adult in their interactions with each other.

Limbic system (often referred to as the emotional brain) is concerned with the registering and storing of emotional information. This includes fear, anger and happiness.

Magnetoencephalography (MEG) is a brain imaging technique that measures the magnetic fields generated by the brain's neuronal activity. During MEG, electrical activity in neurons produces magnetic fields that can be recorded outside the skull and used to calculate the locations of the activity within the brain. Infants and children can be studied using this technique as it is non-invasive.

Mirror neurons are neurons in the brain which are activated both when actions are executed and the actions are observed. They might become activated when an individual observes another carrying out a movement or when observing another

person experiencing an emotion such as happiness, fear or pain.

Motor disorders are disorders of the nervous system that can cause abnormal and involuntary movements. Adults and children often experience associated problems including difficulty in processing information that involve visual and spatial awareness, which is needed to guide their motor actions. This includes having trouble with tasks that require hand-eye coordination.

Myelination is an important developmental process in which the axons of neurons become covered in a white fatty substance, which enables quick information processing. Like changes in synapses, myelination is a process that continues into our twenties.

Neocortex is the most recently evolved part of the cerebral cortex. It is involved in higher functions such as conscious thought, language and generation of motor commands, and it receives and stores information for remembering and decision-making. The neocortex also helps us to judge our responses to our surroundings.

Neuroarcheology refers to the impact of adverse events on the developing brain, with the implicit suggestion that these experiences become indelibly marked upon the brain.

Neuroimaging refers to the production of images of the brain achieved by non-invasive techniques which include functional magnetic resonance imaging (fMRI) and computed tomography (CT scanning).

Neurotransmitters are chemical messengers that carry, boost and control signals between pre-synaptic neurons and other cells in the body. They tend to be released from the neuron's axon terminal after an action potential has reached the synapse.

Oxytocin (sometimes referred to as the 'love hormone') is produced by the pituitary gland and is responsible for human behaviours associated with bonding and relationships. This hormone is greatly stimulated during childbirth, breastfeeding and sex.

Pituitary gland is situated at the base of the brain. It is also known as the 'master' gland because it is often considered the most important part of the endocrine system as it produces hormones that control many functions of other endocrine glands.

Plasticity refers to the brain's unique ability to change its structure and function as a result of changes internally (within the body) or externally (in the environment).

Positron emission tomography (PET scan) is an imaging technique that helps to show how tissues and organs function. A PET scan uses a radioactive drug (tracer) to show this activity. The tracer may be injected, swallowed or inhaled, depending on which organ or tissue is being studied by the PET scan. The tracer collects in areas of the body that have higher levels of chemical activity, which often correspond to areas of disease, which on a PET scan show up as bright spots.

Prefrontal cortex (PFC) is situated behind the forehead in the frontal lobe of the brain and plays an important role in the regulation of complex cognitive, emotional and behavioural functioning. It is also responsible for many higher-level thinking skills such as decision-making.

Primary Health Care Team (PHCT) is a multidisciplinary group of health and social care professionals who work together to deliver local accessible services to a defined population. A PHCT usually consists of a practice manager, doctors, nurses, health visitors and support staff (receptionists, secretaries and clerical staff), as well as midwives.

Schizophrenia is a brain disorder that affects the way a person behaves, thinks and sees the world. Schizophrenia makes it difficult to distinguish between what is real and unreal, think clearly, manage emotions, relate to others and function normally. There is a wide range of care available which includes therapy, medication and support groups.

Social referencing is what children often do to interpret whether situations are safe or pose a threat to their well-being. It involves observing an adult's facial cues, actions, gestures and facial expressions in an attempt to figure out what other people are thinking and feeling, and what they are about to do next.

Spinal cord is situated in the spinal column and connects the brain with nerves that branch out to the rest of the body.

Synapses are the tiny gaps between neurons.

Synaptic pruning is the process that eliminates weaker synaptic connections, while stronger connections are kept and strengthened. Experience determines which connections will be strengthened and which will be pruned.

Bibliography

Abell, F., Krams, M. and Ashburner, J. (1999). 'The Neuroanatomy of Autism: A Voxel-Based Whole Brain Analysis of Structural Scans'. *Cognitive Neuroscience* 10: 1647–51.

Acharya, S. and Shukla, S. (2012). 'Mirror Neurons: Enigma of the Metaphysical Modular Brain'. *Journal of Natural Science, Biology and Medicine* 3 (2): 118–24. DOI: 10.4103/0976-9668.101878.

Action for Children (2015). *Impact Report* (2015).Watford: Action for Children.

Adolphs, R. (2009). 'The Social Brain: Neural Basis of Social Knowledge'. *Annual Review of Psychology* 60: 693–716 (Bethesda, MD: US National Library of Medicine).

Allen, G. (2011). *Early Intervention: The Next Steps*. London: Crown Copyright.

Allen, G. and Duncan Smith, I. (2008). *Early Intervention: Good Parents, Better Kids, Better Citizens*. London: The Centre for Social Justice.

Almond, D. and Currie, J. (2010). 'Human Capital Development before Age Five'. *Handbook of Labour Economics* 4b: 1315–486. New York: Elsevier Ltd.

Ansari, D., Coch, D. and De Smedt, B. (2011). 'Connecting Education and Cognitive Neuroscience: Where Will the Journey Take Us?' *Educational Philosophy Theory* 43: 37–42.

Arnold, J. C. (2014). *Their Name Is Today: Reclaiming Childhood in a Hostile World*. Robertsbridge: Plough Publishing House.

Aslin, R. N. and Mehler, J. (2005). 'Near-infrared Spectroscopy for Functional Studies of Brain Activity in Human Infants: Promise, Prospects and Challenges'. *Journal of Biomedical Optics* 10: 11009 (PubMed).

Athey, C. (1990). *Extending Thought in Young Children*. London: Paul Chapman.

Azevedo, F. A. C., Carvalho, L. R. B., Grinberg, L. T., Farfel, J. M., Ferretti, R. E. L., Leite, R. E. P., Filho, W. J., Lent, R. and Herculano-Houzel, S. (2009). 'Equal Number of Neuronal and Nonneuronal Cells Make the Human Brain an Isometrically Scaled-Up Primate Brain'. *Journal of Comparative Neurology* 513: 532–41. DOI: 10.1002/cne.21974.

Bada, H. S., Das, A., Bauer, C., Shankaran, S., Lester, B. M., Gard, C. C., Wright, L. L., LaGasse, L. and Higgins, R. (2005). 'Low Birth Weight and Preterm Births: Etiologic Fraction Attributable to Prenatal Drug Exposure'. *Journal of Perinatology* 25: 631–7.

Bandura, A. (1965). 'Influence of Models' Reinforcement Contingencies on the Acquisition of Imitative Responses'. *Journal of Personality and Social Psychology* 1 (6): 589–95.

Barker, D. (1995). *Nutrition in the Womb*. London: Random House.

Barrett, L. F., Mesquita, B., Ochsner, K. N. and Gross, J. J. (2007). 'The Experience of Emotion'. *Annual Review of Psychology*. 58: 373–403.

Barnett, W. S., Jung, K., Yarosz, Z., Thomas, J., Hornbeck, A., Stechuk, R. and

Burns, S. (2008). 'Educational Effects of the Tools of the Mind Curriculum: A Randomized Trial'. *Early Childhood Research Quarterly* 23 (3): 299–313.

Baron-Cohen, S. (2004). 'The Cognitive Neuroscience of Autism.' *Journal of Neurology, Neurosurgery and Psychiatry* 75: 945–8.

Barr, R. and Hayne, H. (1999). 'Developmental Changes in Imitation from television During Infancy'. *Child Development* 70: 1067–81.

Barron-Hauwaert, S. (2004). *Language Strategies for Bilingual Families: The One Parent-One Language Approach*. Clevedon: Multilingual Matters.

Bauman M. and Kemper T. (1994). *The Neurobiology of Autism*. Baltimore, MD: Johns Hopkins.

Bazhenova, O. V., Stroganova, T. A., Doussard-Roosevelt, J. A., Posikera, I. A. and Porges, S. W. (2007). 'Physiological Responses of Five-month-old Infants to Smiling and Blank Faces'. *International Journal of Psychophysiology.* 63: 64–76.

Bear, M. F, Connors, B. W. and Paradiso, M. A. (2007). *Neuroscience. Exploring the Brain*. Baltimore, MD: Lippincot Williams and Wilkins.

Bell, M. A. and Wolfe, C. D. (2007). 'Changes in Brain Functioning from Infancy to Early Childhood: Evidence from EEG Power and Coherence During Working Memory Tasks'. *Developmental Neuropsychology.* 31: 21–38.

Bell, M. A. and Wolfe, C. D. (2004). 'Emotion and Cognition: An Intricately Bound Developmental Process'. *Child Development.* 75(2): 366–70.

Bengtsson, S. L., Nagy, Z., Skare, S., Forsman, L., Forssberg, H. and Ullén, F. (2005). 'Extensive Piano Practicing has Regionally Specific Effects on White Matter Development'. *Nature Neuroscience* 8: 1148–50.

Benoit, D. Parker, K. and Zeanah, C. (1997). 'Mothers' Representations of their Infants Assessed Pre-natally: Stability and Association with Infants' Attachment Classifications'. *Journal of Child Psychology, Psychiatry and Allied Disciplines* 38: 307–13.

Bennethan, M. (2009). 'Nurture Groups: Early Relationships and Mental Health. In Cefai, C. and Cooper, P. (eds), *Promoting Emotional Education: Engaging Children and Young People with Social, Emotional and Behavioural Difficulties.* (144–50). London: Jessica Kingsley Publishers.

Berger, A., Tzur, G. and Posner, M. I. (2006). 'Infant Brains Detect Arithmetic Errors'. *Proceedings of the National Academy of Sciences of the United States of America* 103 (33): 12649–53.

Bialystok, E. (2001). *Bilingualism in Development: Language, Literacy and Cognition*. New York: Cambridge University Press.

Bialystok, E. and Senman, L. (2004). 'Executive Processes in Appearance-reality Checks: The Role of Inhibition of Attention and Symbolic Representation'. *Child Development* 75 (2): 562–79.

Bishop, D. V. M. (2013). 'Research Review: Emanuel Miller Memorial Lecture 2012 – Neuroscientific Studies of Intervention for Language Impairment in Children: Interpretive and Methodological Problems'. *Journal of Child Psychology and Psychiatry, and Allied Disciplines* 54(3): 247–59.

Bishop, D. V. M. (2000). 'How Does the Brain Learn Language? Insights from the Study of Children with and without Language Impairment'. *Developmental Medicine and Child Neurology* 42 (2): 133–42.

Bjørkvold, J.-R. (1992). *The Muse Within: Creativity and Communication, Song and Play from Childhood through Maturity*. New York: Harper Collins.

Blakemore, S. J. and Frith, U. (2005). *The Learning Brain: Lessons for Education.* Oxford: Blackwell.

Bodrova, E. and Leong, D. J. (2007). *Tools of the Mind*, 2nd edn. Columbus, OH: Merrill/Prentice Hall.

Bortfeld, H., Wruck, E. and Boas, D. A. (2007). 'Assessing Infants' Cortical Response to Speech Using Near-infrared Spectroscopy'. *NeuroImage* 34: 407–15.

Bosseler, A. N., Taulu, S., Pihko, E., Mäkelä, J. P., Imada, T., Ahonen, A. and Kuhl, P. (2013). 'Theta Brain Rhythms Index Perceptual Narrowing in Infant Speech Perception'. *Frontiers in Psychology* 4 (690): 1–12.

Bowlby, J. (1953). *Child Care and the Growth of Love.* London: Penguin.

Boxall, M. (2002). *Nurture Groups in Schools: Principles and Practice.* London: Sage Publications.

Boyd, J., Barnett, W. S., Bodrova, E., Leong, D. and Gomby, D. (2005). 'Promoting Children's Social and Emotional Development through Preschool Education'. *Preschool Policy Brief.* New Brunswick, NJ: National Institute for Early Education Research.

Bradley, R., Atkinson, M., Tomasino, D. and Rees, R. A. (2009). *Facilitating Emotional Self-Reguation in Preschool Children: Efficacy of the Early HeartSmarts Program in Promoting Social, Emotional and Cognitive Development.* Boulder Creek, CA: HeartMath LLC.

Brazelton, T. B. (1992). *To Listen to a Child: Understanding the Normal Problems of Growing Up.* Cambridge, MA: Perseus.

Brazelton, T. B. and Nugent, J. K. (1995). *The Neonatal Behavioural Assessment Scale.* Cambridge: MacKeith Press.

Briggs-Gowan, M. J., Carter, A. S., Bosson-Heenan, J., Guyer, A. E. and Horwitz, S. M. (2006). 'Are Infant-toddler Socio-emotional and Behavioural Problems Transient?' *Journal of the American Academy of Child and Adolescent Psychiatry* 45: 849–58.

British Heart Foundation (2016). *The Best Start in Life – A Manifesto for Physical Activity in the Early Years.* Leicester: Loughborough University.

Brotherson, S. (2009). *Understanding Brain Development in Young Children.* Bright Beginnings, North Dakota State University (NDSU) Extension Service. Fargo: North Dakota State University.

Bruer, J. (2011). *Revisiting The Myth of the First Three Years.* Canterbury: University of Kent, Centre for Parenting Culture Studies.

Bruner, J. S. (1968). *Processes of Cognitive Growth: Infancy.* Heinz Werner Lectures, 1968. Worcester, MA: Clark University Press with Barri Publishers.

Bruner, J. (1996). *The Culture of Education.* Cambridge, MA: Harvard University Press.

Bruner, J. S. (2003). *Making Stories: Law, Literature, Life.* Cambridge, MA: Harvard University Press.

Bullowa, M. (ed.) (1979). *Before Speech: The Beginning of Human Communication.* London: Cambridge University Press.

Burke, H. M., Davis, M. C., Otte, C. and Mohr, D. C. (2005). 'Depression and Cortisol Responses to Psychological Stress: A Meta-Analysis'. *Psychoneuroendocrinology* 30: 846–56.

Byers-Heinlein, K. and Lew-Williams, C. (2013). *Learning Landscapes* 7 (1). http://www.learninglandscapes.ca/images/documents/ll-no13/byers-heinlein.pdf (accessed 16 June 2016).

Byers-Heinlein, K., Burns, T. C. and Werker, J. F. (2010). 'The Roots of Bilingualism in Newborns'. *Psychological Science* 21 (3): 343–8.

Chasnoff, I., and MacGregor, S. (1987). 'Maternal Cocaine Use and Neonatal Morbidity'. *Paediatric Research* 21: 356A.

Chomsky, N. (1965). *Aspects of the Theory of Syntax. 50th Anniversary Edition.* London: MIT Press.

Christakis, D. A. (2009). 'The Effects of Infant Media Usage: What Do We Know and What Should we Learn?' *Acta Paediatrica* 98 (1): 8–16.

Chugani, H. (1998). 'A Critical Period of Brain Development: Studies of Cerebral Glucose Utilization with PET'. *Preventive Medicine* 27: 184–8.

Chugani, H. T., Behen, M. E., Muzik, O., Juhasz, C. Nagy, F. and Chugani, D. C. (2001). 'Local Brain Functional Activity Following Early Deprivation: A Study of Post-institutionalized Romanian Orphans'. *Neuroimage* 14 (6): 1290–301.

Chukovsky K. (1968). *From Two to Five*. Berkley: University of California Press.

Cicchetti, D. and Tucker, D. (1994). 'Development and Self-Regulatory Structures of the Mind'. *Development and Psychopathology* 6: 533–49.

Clark, C. and Dugdale, G. (2008). *Literacy Changes Lives: The Role of Literacy in Offending Behaviour*. London: National Literacy Trust.

Cohen, J., Onunaku, N., Clothier, S. and Poppe, J. (2005). *Helping Young Children Succeed: Strategies to Promote Early Childhood Social and Emotional Development*. Washington, DC: National Conference of State Legislatures and Zero to Three.

Coleridge, S. T. (1802). Cited in Coleridge, H. (1836). *The Literary Remains of Samuel Taylor Coleridge: Volume 1*. London: William Pickering.

Coltheart, M. and McArthur, G. (2012). 'Neuroscience, Education and Educational Efficacy Research'. In Anderson, M. and Della Sala, S. (eds), *Neuroscience in education*. Oxford: Oxford University Press, 215–21.

Comeau, L., Genesee, F. and Lapaquette, L. (2003). 'The Modelling Hypothesis and Child Bilingual Codemixing'. *International Journal of Bilingualism* 7 (2): 113–26.

Commodari, E. (2013). 'Preschool Teacher Attachment School Readiness and Risk of Learning Difficulties'. *Early Childhood Research Quarterly* 28: 123–33.

Conti, G., Heckman, J. and Urzua, S. (2011). *Early Endowments, Education and Health: Human Capital and Economic Opportunity. Working Paper 2011–001*. Chicago: University of Chicago.

Cooper, P. and Tiknaz, Y. (2007). *Nurture Groups in School and at Home: Connecting with Children with Social, Emotional and Behavioural Difficulties*. London: Jessica Kingsley.

Corel, J. L. (1975). *The Postnatal Development of the Human Cerebral Cortex*. Cambridge, MA: Harvard University Press.

Cozolino, L. (2013). *The Social Neuroscience of Education: Optimizing Learning and Attachment in the Classroom*. London: W. W. Norton and Co.

Cunha, F., Heckman, J. J., Lochner, L. J. and Masterov, D. V. (2006). 'Interpreting the Evidence on Life Cycle Skill Formation'. In Hanushek, E. A. and Welch, F. (eds), *Handbook of the Economics of Education* 12. Amsterdam: North-Holland, 697–81.

Dapretto, M., Davies, M. S., Pfeifer, J. H., Scott, A. A., Sigman, M., Bookheimer, S. Y. and Marco Iacoboni, M. (2006). 'Understanding Emotions in Others:

Mirror Neuron Dysfunction in Children with Autism Spectrum Disorders'. *Nature Neuroscience* 9: 28–30.

Darwin, C. (1872). *The Expressions of Emotion in Man and Animals*. New York: D. Appleton.

David, T., Goouch, K., Powell, S. and Abbott, L. (2003). *Young Brains*. Research Report Number 444. London: Department for Education and Skills.

Davidson, R. and Fox, N. (1992). 'Assymetrical Brain Activity Discriminates between Positive v. Negative Stimuli in Human Infants'. *Science* 218: 1235–7.

Davis, O. S. P., Haworth, C. M. A., Lewis, C. M. and Plomin, R. (2012). 'Visual Analysis of Geocoded Twin Data puts Nature and Nurture on the Map'. *Molecular Psychiatry* 17: 867–74. DOI: 10.1038/mp.2012.68.

Daycare Trust (2010). *Supporting Parents in Helping their Children to Learn at Home: Some Tips for Childcare Providers*. London: Daycare Trust.

De Bellis, M. D. and Kuchibhatla, M. (2006). 'Cerebellar Volumes in Paediatric Maltreatment-related Posttraumatic Stress Disorder'. *Biological Psychiatry* 60: 697–703.

DeCasper, A. J. and Prescott, P. (2009). 'Lateralised Processes Constrain Auditory Reinforcement in Human New-borns'. *Hearing Research* 255: 135–41.

DeCasper, A. J. and Spence, M. J. (1986). 'Prenatal Maternal Speech Influences Newborns' Perception of Speech Sounds'. *Infant Behaviour and Development* 9 (2): 133–50.

DeCasper, A. J. and Spence, M. J. (2006). 'Prenatal Maternal Speech Influences New-borns' Perception of Speech Sounds'. *Infant Behaviour and Development* 9: 133–50.

Decety, J. and Jackson, P. L. (2004). 'The Functional Architecture of Human Empathy'. *Behavioural and Cognitive Neuroscience Reviews Journal* 3: 71–100.

Dekker, S., Lee, N. C., Howard-Jones, P. and Jolles, J. (2012). 'Neuromyths in Education: Prevalence and Predictors of Misconceptions among Teachers'. *Front. Psychology* 3: 429. DOI: 10.3389/fpsyg.2012.00429.

Delafield-Butt, J. T. and Gangopadhyay, N. (2013). 'Sensorimotor Intentionality: The origins of Intentionality in Prospective Agent Action'. *Developmental Review* 33: 399–425. DOI: 10.1016/j.dr.2013.09.001.

Delafield-Butt, J. and Trevarthen, C. (2013). 'Theories of the Development of Human Communication'. In Cobley, P. and Schultz, P. J. (eds), *Handbook of Communication Science, Vol. 1: Theories and Models of Communication*. Berlin: De Gruyter Mouton, 199–221.

Denham, S. and Kochanoff, A. T. (2002). 'Parental contributions to Pre-schoolers' Understanding of Emotion'. *Marriage and Family Review* 34 (3–4): 311–43.

Department for Children, Schools and Families (DCSF) (2008). *The Bercow Report: A Review of Services for Children and Young People (0–19) with Speech, Language and Communication Needs*. Nottingham: Department for Children, Schools and Families.

Department for Education (2011). *The Report of the Independent Review on Poverty and Life Chances*. London: Department for Education.

Department for Education and the Department of Health (2015). *Promoting the Health and Well-being of Looked-after Children: Statutory Guidance for Local Authorities, Clinical Commissioning Groups and National Health Service England*. London: Department for Education and the Department of Health.

Department for Education and Skills, (DfES) (2014). *The Early Years Foundation Stage: Statutory Framework for the Early Years Foundation Stage: Setting the Standards for Learning, Development and Care for Children from Birth to Five*. London: Department for Education and Skills.

Depue, R., Luciana, M., Arbisi, P., Collins, P. and Leon, A. (1994). 'Dopamine and the Structure of Personality: Relation of Agonist Induced Dopamine Activity to Positive Emotionality'. *Journal of Personality and Social Psychology* 67: 485–98.

DeRosnay, M., Cooper, P. J., Tsigaras, N. and Murray, L. (2006). 'Transmission of Social Anxiety from Mother to Infant: An Experimental Study Using a Social Referencing Paradigm'. *Behaviour Research and Therapy* 44 (8): 1165–75.

Desforges, C. and Abouchaar, A. (2003). *The Impact of Parental Involvement, Parental Support and Family Education on Pupil Achievement and Adjustment: A Literature Review*. London: DfES.

Detillion, C. E., Craft, T. K., Glasper, E. R., Prendergast, B. J. and DeVries, A. C. (2004). 'Social Facilitation of Wound Healing'. *Psychoneuroendocrinology* 29 (8): 1004–11.

Devonshire, I. M. and Dommett, E. J. (2010). 'Neuroscience: Viable Applications in Education?' *Neuroscientist* 16: 349–56.

Diamond, A., Barnett, W. S., Thomas, J. and Munro, S. (2007). 'Preschool Programme Improves Cognitive Control'. *Science* 318: 1387–8.

Donaldson, M. (1978). *Children's Minds*. Glasgow: Fontana/Collins.

Donaldson, M. (1992). *Human Minds: An Exploration*. London: Allen Lane/ Penguin Books.

Dowling, J. E. (2004). *The Great Brain Debate*. Princeton and Oxford: Princeton University Press.

Early Intervention Foundation (2015). *The Best Start at Home. A Report on What Works to Improve the Quality of Parent Child Interactions from Conception to Age Five*. London: Early Intervention Foundation.

Eliot, L. (1999). *What's Going On in There? How the Brain and Mind Develop in the First Five Years*. New York: Bantam Books.

Eluvathingal, T., Chugani, H., Behen, M., Juhasz, C., Muzik, O. and Maqbool, M. (2006). 'Abnormal Brain Connectivity in Children after Early Severe Socioemotional Deprivation: A Diffusion Tensor Imaging Study'. *Paediatrics* 117 (6): 2093–100.

Erikson, E. (1950). *Childhood and Society*. London: Pelican Books.

Family and Parenting Institute (2009). *Early Home Learning Matters: A Brief Guide for Practitioners*. London: Family and Parenting Institute.

Feinstein, L. (2000). *The Relative Economic Importance of Academic, Psychological and Behavioural Attributes Developed in Childhood*. Centre for Economic Performance Discussion Paper 443. London: London School of Economics and Political Science.

Field, F. (2010). *The Foundation Years: Preventing Poor Children Becoming Poor Adults*. London: Crown Copyright.

Fields, R. D. (2008). 'White Matter in Learning, Cognition and Psychiatric Disorders'. *Trends in Neurosciences* 31 (7): 361–70.

Fletcher, P. C., Happe, F., Frith, U., Baker, S. C., Dolan, R. J., Frackowiak, R. S. J. and Frith, C. D. (1995). 'Other Minds in the Brain: A Functional

Imaging Study of "Theory of Mind" in Story Comprehension'. *Cognition* 57: 109–28.

Florio, M. and Huttner, W. B. (2014). 'Neural Progenitors, Neurogenesis and the Evolution of the Neocortex'. *Development* 141: 2182–94. DOI: 10. 1042/dev. 090571 (Cambridge: Company of Biologists Ltd).

Fonagy, P. (1994). 'The Theory and Practice of Resilience'. *Journal of Child Psychology and Psychiatry* 35 (2): 231–57.

Fonagy, P. (2001). *Attachment Theory and Psychoanalysis*. London: Karnac Books Ltd.

Fonagy, P., Gergely, G., Jurist, E. L. and Target, M. (2002). *Affect Regulation, Mentalization and the Development of the Self*. New York: Other Press.

Fonteneau, E. and Lely, H. K. J. van der (2008). 'Electrical Brain Responses in Language Impaired Children Reveal Grammar-Specific Deficits'. *PLoS ONE* 3 (3): e1832. DOI: 10.1371/journal.pone.0001832.

Fox, N. A. and Schonkoff, J. P. (2011). 'Violence and Development: How Persistent Fear and Anxiety can Affect Young Children's Learning, Behaviour and Health'. In Bernard van Leer Foundation (ed.), *Hidden Violence: Protecting Young Children at Home*. Early Childhood Matters No. 116. The Hague: Bernard van Leer Foundation.

Fried, P. A. and Watkinson, B. (2001). 'Differential Effects on Facets of Attention in Adolescents Prenatally Exposed to Cigarettes and Marijuana'. *Neurotoxicol Teratol* 23: 421–30.

Frith, C. (2007). 'The Social Brain?' *Proceedings of the Royal Society B: Biological Sciences* 362 (1480): 671–8.

Frith, U. (2003). *Autism: Explaining the Enigma*. Oxford: Blackwell Publishers.

Frith, U. and Happe, F. (1999). 'Theory of Mind and Self-Consciousness: What Is It Like To Be Autistic?' *Mind and Language* 14 (1): 1–22 (Oxford: Blackwell Ltd).

Gallese, V. (2001). 'The Shared Manifold Hypothesis: From Mirror Neurons to Empathy'. *Journal of Consciousness Studies* 8: 33–50.

Gallese, V. (2005). 'Embodied Simulation: From Neurons to Phenomenal Experience'. *Phenomenology and the Cognitive Sciences* 4: 23–48.

Gazzola, V. and Keysers, C. (2009). 'The Observation and Execution of Actions Share Motor and Somatosensory Voxels in all Tested Subjects: Single-subject Analyses of Unsmoothed fMRI Data'. *Cerebral Cortex* 19: 1239–55.

Geddes, H. (2006). *Attachment in the Classroom: The Links between Children's Early Experiences, Emotional Well-being and Performance in School*. London: Worth Publishing.

Gerhardt, S. (2015). *Why Love Matters*. London and New York: Routledge.

Glick, T., Livesey, S. J. and Wallis, F. (2005). *Medieval Science, Technology, and Medicine: An Encyclopedia*. Abingdon: Taylor and Francis Ltd.

Goetz, P. J. (2003). 'The Effects of Bilingualism on Theory of Mind Development'. *Bilingualism: Language and Cognition* 6 (1): 1–15.

Goldacre, B. (2009). *Bad Science*. London: Fourth Estate.

Goldstein, M. (1994). 'Neurologic Effects of Alcoholism'. *Western Journal of Medicine* 161 (3): 279–87.

Goleman, D. (1996). *Emotional Intelligence*. Cambridge, MA: Harvard University Press.

Goodrich, B. G. (2010). 'We Do, Therefore We Think: Time, Motility, and Consciousness'. *Reviews in the Neurosciences* 21: 331–61.

Gopnik, A. (2009). *The Philosophical Ba*by. London: Bodley Head.

Gopnik, A., Meltzoff, A. and Kuhl, P. (1999). *The Scientist in the Crib: What Early Learning Tells Us About the Mind*. New York: HarperCollins.

Goswami, U. (2006). 'Neuroscience and Education: From Research to Practice? Nature Reviews'. *Neuroscience* 7: 406–13.

Goswami, U. (2015). *Children's Cognitive Development and Learning*. Cambridge: Cambridge Primary Review Trust.

Gottmann, J. M., Katz, L. F. and Hooven, C. (1996). 'Parental Meta-Emotion Philosophy and the Emotional Life of Families: Theoretical Models and Preliminary Data'. *Journal of Family Psychology* 10 (1): 243–68.

Gray, J. (2009). *Why Mars and Venus Collide*. New York: HarperCollins.

Greenspan, R. J. (1995). 'Understanding the Genetic Construction of Behaviour'. *Scientific American* 272 (4) (April): 72–8.

Grice, S. J., Halit, H., Farroni, T., Baron-Cohen, S., Bolton P. and Johnson, M. H. (2005). 'Neural Correlates of Eye-gaze Detection in Young Children with Autism'. *Cortex* 41: 342–53.

Gunnar M. and Davis, E. P. (2003). 'The Developmental Psychobiology of Stress and Emotion in Early Childhood'. In Weiner, I. B., Lerner, R. M., Easterbrooks, M. A. and Mistry, J. (eds), *Comprehensive Handbook of Psychology*, Vol. 6, *Developmental Psychology*. New York: Wiley, 113–43.

Gunnar, M. and Quevado, K. (2007). 'The Neurobiology of Stress and Development'. *Annual Review of Psychology* 58: 145–73 (Bethesda, MD: US National Library of Medicine).

Gunnar, M. R. and Vazquez, D. (2006). 'Stress Neurobiology and Developmental Psychopathology'. In Cicchetti D. and Cohen D. (eds), *Developmental Psychopathology*, Vol. 2: *Developmental Neuroscience*, 2nd edn. New York: Wiley.

Gutman, L. and Feinstein, L. (2007). *Parenting Behaviours and Children's Development from Infancy to Early Childhood: Changes, Continuities, and Contributions*. London: Centre for Research on the Wider Benefits of Learning.

Hackman, A. D. and Farah, J. M. (2009). 'Socio-economic Status and the Developing Brain'. *Trends in Cognitive Science* 13 (2): 65–73.

Hamilton, A. and Marsh, L. (2013). 'Two Systems for Action Comprehension in Autism'. In Baron-Cohen, S., Tager-Fusberg, H. and Lombardo, M. V. (eds), *Understanding Other Minds: Perspectives from Developmental Social Neuroscience*. Oxford: Oxford University Press.

Happe, F. and Frith, U. (1994). 'Theory of Mind in Autism'. In Schopler, E. and Mesibov, G. B. (eds), *Learning and Cognition in Autism*. New York: Plenum.

Harlow, H. F. (1958). 'The Nature of Love'. *American Psychologist* 13 (12): 673–85. DOI: 10.1037/h0047884.

Hayne, H., Herbert, J. and Simcock, G. (2003). 'Imitation from Television by 24- and 30-month-olds'. *Developmental Science* 6: 254–61.

Helsper, E. J., Kalmus, V., Hasebrink, U., Sagvari, B. and De Haan, J. (2013). 'Country Classification: Opportunities, Risks, Harm and Parental Mediation'. http://eprints.lse.ac.uk/52023/ (accessed 23 September 2015).

Herculano-Houzel, S. (2009). 'The Human Brain in Numbers: A Linearly Scaled-up Primate Brain'. *Frontiers in Human Neuroscience* 3: 31. DOI: 10.3389/neuro.

Hickock, G. (2009). 'Eight Problems for the Mirror Neuron Theory of Action

Understanding in Monkeys and Humans'. *Journal of Cognitive Neuroscience* 21 (7): 1229–43.

Hofsten, C. von (2007). 'Action in Development'. *Developmental Science* 10: 54–60. DOI: 10.1111/j.1467-7687.2007.00564.x.

Holloway, D., Green, L. and Livingstone, S. (2013). *Zero to Eight: Young Children and Their Internet Use.* London: LSE, EU Kids Online.

Howard-Jones, P. (2007). *Neuroscience and Education: Issues and Opportunities.* London: TLRP, Institute of Education.

Howard-Jones, P. (2010). *Introducing Neuroeducational Research: Neuroscience, Education and the Brain from Contexts to Practice.* Abingdon: Routledge.

Howard-Jones, P. (2014). Neuroscience and Education: *A Review of Educational Interventions and Approaches Informed by Neuroscience.* London: Education Endowment Foundation.

Howard-Jones, P. (2015). Personal written communication.

Howard-Jones, P., Franey, L., Mashmoushi, R. and Liao, Y. C. (2009). *The Neuroscience Literacy of Trainee Teachers.* Leeds: Education – Line.

Howard-Jones, P., Pickering, S. and Diack, A. (2007). *Perceptions of the Role of Neuroscience in Education.* London: Innovation Unit.

Hubel, D. H. and Wiesel, T. N. (1962). 'Receptive Fields, Binocular Interaction and Functional Architecture in the Cat's Visual Cortex'. *The Journal of Physiology* 160 (1): 106–54.

Hughes, D. A. and Baylin, J. (2012). *Brain-based Parenting: The Neuroscience of Caregiving for Healthy Attachment.* London and New York: W. W. Norton and Co.

Huth, A. G. (2016). 'Detailed Map of Language Representation in Human Brain'. http://neuroscience.berkeley.edu/2477-2/ (accessed 17 June 2016).

Huth, A. G., Heer, W. A. de, Griffiths, T. L., Theunissen, F. E. and Gallant, J. L. (2016). 'Natural Speech Reveals the Semantic Maps that Tile Human Cerebral Cortex'. *Nature* 532: 453–8.

Iacoboni, M. (2012). Cited in Frank, L. (2012). *The Neurotourist: Postcards from the Edge of Brain Science.* Oxford: One World.

ICAN (2006). *I CAN Cost to the Nation of Children's Poor Communication Report.* Issue 2. London: I CAN.

Immordino-Yang, M. H. and Damasio, A. (2007). 'We Feel, Therefore we Learn: The Relevance of Affective and Social Neuroscience to Education'. *Mind, Brain and Education Journal* 1 (1): 3–10.

Independent Regulator and Competition Authority for the UK Communications Industries (Ofcom) (2015). *Children and Parents: Media Use and Attitudes Report.* London: Ofcom.

Iversen, J. M. (2010). 'Developing Language in a Developing Body: The Relationship between Motor Development and Language Development'. *Journal of Child Language* 37: 229–61.

Jiang, J., Dai, B., Peng, D., Zhu, C., Liu, L. and Lu, C. (2012). 'Neural Synchronization during Face-to-face Communication'. *Journal of Neuroscience* 32: 16064–9.

Jolles, J., De Groot, R. H. M., Van Benthem, J., Dekkers, H., De Glopper, C. and Uijlings, H. (2005). *Brain Lessons.* Maastricht: Neuropsych.

Jones, W., Carr, K. and Klin, A. (2008). 'Absence of Preferential Looking to the Eyes of Approaching Adults Predicts Level of Social Disability in 2-year-olds with Autism Spectrum Disorder'. *Archives of General Psychiatry* 65: 946–54.

Karmilloff-Smith, A. and Karmilloff, K. (2015). *Understanding Your Baby: A Parent's Guide to Progress in the First Year and Beyond.* London: Hamlyn.

Kaufman, J. and Charney, D. (2001). 'Effects of Early Stress on Brain Structure and Function: Implications for Understanding the Relationship between Child Maltreatment and Depression'. *Development and Psychopathology* 13: 451–71.

Kinlein, S. A., Wilson, C. D. and Karatsoreos, I. N. (2015). 'Dysregulated Hypothalamic–Pituitary–Adrenal Axis Function Contributes to Altered Endocrine and Neurobehavioral Responses to Acute Stress'. *Frontiers in Psychiatry* 6: 31.

Kisilevsky, B. S., Hains, S. M. J., Brown, C. A., Lee, C. T., Cowperthwaite, B., Stutzman, M. L., Swansburg, K., Xie, X., Huang, H., Ye, H., Hzng, K. and Wang, Z. (2008). 'Foetal Sensitivity to Properties of Maternal Speech and Language'. *Infant Behaviour and Development* 32: 59–71.

Klin, A., Klaiman, C. and Jones, W. (2015). 'Reducing Age of Autism Diagnosis: Developmental Social Neuroscience Meets Public Health Challenge'. *Revue Neurologique* 60: S3–S11.

Koizumi, H. (2004). 'The Concept of Developing the Brain: A New Natural Science for Learning and Education'. *Brain Development* 6 (7): 434–41.

Kolb, B. (2009). 'Brain and Behavioural Plasticity in the Developing Brain: Neuroscience and Public Policy'. *Paediatrics and Child Health* 14 (10): 651–2.

Kolata, G. (2007). 'A surprising secret to a long life: Stay in school'. *New York Times*, 3 January 2007. http://www.nytimes.com/2007/01/03/health/03aging.html. (accessed 12 March 2016).

Kovács, A. M. (2009). 'Early Bilingualism Enhances Mechanisms of False-belief Reasoning'. *Developmental Science* 12 (1): 48–54.

Kuhl, P. (2011) 'Early Language Learning and Literacy: Neuroscience Implications for Education'. *Mind, Brain and Education* 5 (3): 128–42.

Kuhl, P. (2010). 'Brain Mechanisms in Early Language Acquisition'. *Neuron* 67 (5): 713–27.

Kuhl, P. (2004). 'Early Language Acquisition: Cracking the Speech Code'. Nature Reviews Neuroscience 5 (11): 831–43.

Kuhl, P. K., Stevens, E., Hayashi, A., Deguchi, T., Kiritani, S. and Iverson, P. (2006). 'Infants Show Facilitation for Native Language Phonetic Perception between 6 and 12 months'. *Developmental Science* 9: 13–21.

Laible, D. (2004). 'Mother–child Discourse Surrounding a Child's Past Behaviour at 30 months: Links to Emotional Understanding and Early Conscious Development at 36 months'. *Merrill-Palmer Quarterly* 50 (2): 159–80.

Lake, A. and Chan, M. (2014). *Putting Science into Practice for Early Child Development.* Published online: 19 September 2014. Amsterdam: World Health Organization, Elsevier Ltd/Inc/BV (accessed 23 February 2015).

Lambe, M., Hultman, C., Torrang, A., Maccabe, J. and Cnattingius, S. (2006). 'Maternal Smoking During Pregnancy and School Performance at Age 15'. *Epidemiology* 17: 524–30.

Langley, K., Rice, F., Bree, M. B. van den and Thapar, A. (2005). 'Maternal Smoking During Pregnancy as an Environmental Risk Factor for Attention Deficit Hyperactivity Disorder Behaviour. A Review'. *Minerva Pediatrica* 57: 359–71.

Lanza, E. (2004). *Language Mixing in Infant Bilingualism: A Sociolinguistic Perspective*. Oxford: Oxford University Press.

Larson, M., White, B. P., Cochran, A., Donzella, B. and Gunnar, M. (1998). 'Dampening of the Cortisol Response to Handling at 3-months in Human Infants and its Relation to Sleep, Circadian Cortisol Activity, and Behavioural Distress'. *Development and Psychobiology* 33: 327–37.

Lashley, K. S. (1951). 'The Problems of Serial Order in Behaviour'. In Jeffress, L. A. (ed.), *Cerebral Mechanisms in Behaviour*. New York: Wiley.

LeDoux, J. (1991). *The Emotional Brain: The Mysterious Underpinnings of Emotional Life*. New York: Simon and Schuster.

LeDoux, J. (2003). *Synaptic Self*. London: Penguin.

Lemma, A. (2010). 'The Power of Relationship: A Study of Key Working as an Intervention with Traumatised Young People'. *Journal of Social Work Practice* 24 (4): 409–27.

Leonard, L. B. (2014). *Children with Specific Language Impairment*. Cambridge, MA: MIT Press.

Lester, B. and Sparrow, J. D. (eds) (2012). *Nurturing Young Children and Their Families: Building on the Legacy of T. B. Brazelton*. Oxford: Wiley-Blackwell Scientific.

Levin, F. M. (1997). 'Integrating Some Mind and Brain Views of Transference: The Phenomena'. *Journal of the American Psychoanalytic Association* 45: 1121–51.

Levitt, P. (1998). 'Prenatal Effects of Drugs of Abuse on Brain Development'. *Drug and Alcohol Dependence* 51: 109–25.

Lewis, T. M. D., Amini, F. M. D. and Lannon, R. M. D. (2001). *A General Theory of Love*. New York: Random House.

Lillycrop, K. A., Hanson, M. A. and Burdge, G. C. (2009). 'Epigenetics and the Influence of Maternal Diet'. In Newnham, J. P. and Ross, M. G. (eds), *Early Life Origins of Human Health and Disease*. Basel, Switzerland: Karger Publishers, 11–20.

Linderkamp, O., Janus, L., Linder, R. and Skoruppa, D. B. (2009). 'Timetable of Normal Foetal Brain Development'. *International Journal of Prenatal and Perinatal Psychology and Medicine* 21 (1/2): 4–16.

Livingstone, S., Haddon, L., Görzig, A. and Ólafsson, K. (2011). *EU Kids Online Final Report*. http://eprints.lse.ac.uk/39351/ (accessed 15 September 2015).

Lloyd-Fox, S., Blasi, A., Volein, A., Everdell, N., Elwell, C. E. and Johnson, M. H. (2009). 'Social Perception in Infancy: A Near Infrared Spectroscopy Study'. *Child Development* 80: 986–99.

Locke, J. (1690/1722). *An Essay Concerning Human Understanding. Book 1 – Innate Notions*. London: Taylor.

Makin, L. and Whitehead, M. (2004). *How to Develop Children's Early Literacy*. London: Paul Chapman.

Malloch, S. and Trevarthen, C. (eds) (2010). *Communicative Musicality: Exploring the Basis of Human Companionship*. Oxford: Oxford University Press.

Marmot Review (2010). *Fair Society, Healthy Lives: A Strategic Review of Health Inequalities in England Post-2010*. London: Marmot Review.

Mason, L. (2009). 'Bridging Education and Neuroscience: A Two-way Path is Possible'. *Cortex* 45: 548–9.

Maturana, H., Mpodozis, J. and C. Letelier, J. (1995). 'Brain, language and the origin of human mental functions'. *Biological Research* 28 (1): 15–26.

Mazoyer, B. M., Tzourio, N., Frak, V., Syrota, A., Murayama, N., Levrier, O.,
    Salamon, G., Dehaene, S., Cohen, L. and Mehler, J. (1993). 'The Cortical
    Representation of Speech'. *Journal of Cognitive Neuroscience* 5 (4):
    467–79.
McCabe, D. P. and Castel, A. D. (2008). 'Seeing is Believing: The Effect of Brain
    Images on Judgments of Scientific Reasoning'. *Cognition* 107: 343–52.
McGilchrist, I. (2009). *The Master and His Emissary*. New Haven, CT: Yale
    University Press.
McGowan, P., Sasaki, A., Huang, T., Unterberger, A., Suderman, M., Ernst, C.,
    Meaney, M., Turecki, G. and Szyf, M. (2008). 'Promoter-wide hypermethylation
    of the ribosomal RNA gene promoter in the suicide brain'. *PLoS One* 3: e2085
    10.1371.
McGowan, P. O., Sasaki, A., D'Alessio, A. C., Dymov, S., Labonté, B.,
    Szyf, M., Turecki, G. and Meaney, M. J. (2009). 'Epigenetic Regulation of the
    Glucocorticoid Receptor in Human Brain Associates with Childhood Abuse'.
    Nature Neuroscience 12 (3): 342–8.
Meadows, S. (2016). *The Science Inside the Child*. Oxford: Routledge.
Meaney, M. (2001). 'Maternal Care, Gene Expression and the Transmission of
    Individual Differences in Stress Reactivity across Generations'. *Annual Review
    of Neuroscience* 24: 1161–92 (Bethesda, MD: US National Library of Medicine).
Mehler, J., Jusczyk, P. W., Lambertz, G., Halsted, N., Bertoncini, J. and Amiel-
    Tison, C. (1988). 'A Precursor of Language Acquisition in Young Infants'.
    Cognition 29 (2): 143–78.
Meltzoff, A. N. (1988). 'Imitation of Televised Models by Infants'. *Child
    Development* 59: 1221–9.
Meltzoff, A. N., Kuhl, P. K., Movellan, J. and Sejnowski T. (2009). 'Foundations
    for a New Science of Learning'. *Science* 17: 284–8.
Messler, D. J. and Frawley, M. G. (1994). *Treating the Adult Survivor of Childhood
    Sexual Abuse: A Psychoanalytic Perspective*. New York: Basic Books.
Modell, A. H. (2003). *Imagination and the Meaningful Brain*. Cambridge, MA:
    MIT Press.
Morales, M., Mundy, P. and Rojas, J. (1998). 'Following the Direction of Gaze and
    Language Development in 6-month-olds'. *Infant Behaviour and Development*
    21: 373–7.
Nagy, E. (2011). 'The Newborn Infant: A Missing Stage in Developmental
    psychology'. *Infant Child Development* 20: 3–19. DOI: 10.1002/icd.683.
Nagy, E. and Molnar, P. (2004). 'Homo Imitans or Homo Provocans? Human
    Imprinting Model of Neonatal Imitation'. *Infant Behaviour and Development* 27:
    54–63.
National Scientific Council on the Developing Child (2010b). *Persistent Fear
    and Anxiety Can Affect Young Children's Learning and Development*. Working
    Paper 9. http:// developingchild.harvard.edu/index.php/resources/reports_and_
    working_papers/working_papers/wp9/ (accessed 3 April 2015).
National Scientific Council on the Developing Child (2011). *Children's Emotional
    Development is Built into the Architecture of Their Brains. Working Paper 2.*
    Cambridge, MA: Centre on the Developing Harvard University.
Newman, L., Sivaratnam, C. and Angela Komiti, A. (2015). 'Attachment and
    Early Brain Development – Neuroprotective Interventions in Infant–Caregiver
    therapy'. *Translational Developmental Psychiatry* 3: 28647.

New Scientist (2016). *If at first you don't succeed.* London: New Scientist.

Oates, J., Karmiloff-Smith, A. and Johnson, M. H. (2012). *Early Childhood in Focus 7: Developing Brains.* Milton Keynes: Open University.

O'Connor, T. G., Ben-Shlomo, Y., Heron, J., Golding, J., Adams, D. and Glover, V. (2005). 'Prenatal Anxiety Predicts Individual Differences in Cortisol in Pre-adolescent Children'. *Biological Psychiatry* 58: 211–17.

Office for Standards in Education, Children's Services and Skills (Ofsted) Report Summary (2011). *Supporting Children with Challenging Behaviour Through a Nurture Group Approach.* London: Ofsted.

O'Muircheartaigh, J., Dean, D. C., Dirks, H., Waskiewicz, N., Lehman, K. and Jerskey, B. A. (2013). 'Interactions between White Matter Asymmetry and Language during Neurodevelopment'. *Journal of Neuroscience* 33 (41): 16170–7.

Orekhova, E. V., Stroganova, T. A. and Posikera, I. N. (1999). 'Theta Synchronization during Sustained Anticipatory Attention in Infants over the second Half of the First Year of Life'. *International Journal of Psychophysiology* 32: 151–72.

Organisation for Economic Cooperation and Development (OECD) (2007). *Understanding the Brain: The Birth of a Learning Science.* Paris: OECD.

O'Sullivan, J. and Chambers, S. (2013): *The Twoness of Twos: Leadership for Two Year Olds.* London: London Early Years Foundation (LEYF).

Page, M. P. A. (2006). 'What Can't Neuroimaging Tell the Cognitive Psychologist?' *Cortex* 42 (3): 428–43. DOI: 10.1016/s0010-9452(08)70375-7.

Panksepp, J. and Biven, L. (2012). *Archaeology of Mind: Neuroevolutionary Origins of Human Emotions.* New York: W. W. Norton and Co.

Paterson, C. (2011). *Parenting Matters: Early Years and Social Mobility.* London: CentreForum.

Pearson, B. Z. (2008). *Raising a Bilingual Child.* New York: Random House.

Percaccio, C. R., Padden, D. M., Edwards, E. and Kuhl, P. K. (2010). 'Native and Nonnative Speech-evoked Responses in High-risk Infant Siblings'. *Abstracts of the International Meeting for Autism Research (IMFAR).* Philadelphia: IMFAR.

Perry, B. D. (2006). 'Applying Principles of Neurodevelopment to Clinical Work with Maltreated and Traumatized Children: The Neurosequential Model of Therapeutics'. In Webb, N. B. (ed.), *Working with Traumatized Youth in Child Welfare.* New York: Guilford Press, 27–52.

Pettus, M. (2006). *Change Your Mind – It's All in Your Head.* Herndon, VA: Capital Books, Inc.

Phelps, E. A. and LeDoux, J. (2005). 'Contributions of the Amygdala to Emotion Processing: From Animal Models to Human Behaviour'. *Neuron* 48 (2): 175–87.

Piaget, J. (1964). *The Early Growth of Logic in the Child.* London: Routledge and Kegan Paul.

Pinker, S. (1995). *The New Science of Language and Mind.* London: Penguin.

Pinker, S. (2007). *The Language Instinct: The New Science of Language and Mind.* London: Folio Society.

Pinker S. and Jackendoff, R. (2005). 'The Faculty of Language: What's Special About It?' *Cognition* 95: 201–36.

Porges, S. W. (1995). 'Orienting in a Defensive World: Mammalian Modifications of Our Evolutionary Heritage. A Polyvagal Theory'. *Psychophysiology* 32: 301–18.

Porges, S. W. (2007). 'The Polyvagal Perspective'. *Biological Psychology* 74 (2): 116–43.

Poulin-Dubois, D., Blaye, A., Coutya, J. and Bialystok, E. (2011). 'The Effects of Bilingualism on Toddlers' Executive Functioning'. *Journal of Experimental Child Psychology* 108 (3): 567–79.

Prado, E. L. and Dewey, K. G. (2014). 'Nutrition and Brain Development in Early Life'. *Nutrition Reviews* 72 (4): 267–284. DOI: 10.1111/nure121.02.

Prechtl, H. F. R. (2001). 'Prenatal and Early Postnatal Development of Human Motor Behaviour'. In Kalverboer, A. F. and Gramsbergen, A. (eds), *Handbook on Brain and Behaviour in Human Development*. Dordrecht: Kluwer Academic Publishers, 415–27.

Quick, R. H. (1894). *Essays on Educational Reformers*. London: Longmans, Green and Company.

Ramachandran, V. S. (2010). *The Tell-Tale Brain: Unlocking the Mystery of Human Nature*. London: Cornerstone.

Read, J., Perry, B. D., Moskowith, A. and Connolloy, J. (2001). 'The Contribution of Early Traumatic Events to Schizophrenia in Some Patients: A Traumagenic Neurodevelopmental Model'. *Psychiatry* 64: 319–45.

Redcay, E. (2008). 'The Superior Temporal Sulcus Performs A Common Function for Social and Speech Perception: Implications for the Emergence of Autism'. *Neuroscience and Biobehavioural Reviews* 32: 123–42 (PubMed).

Reddy, V. (2008). *How Infants Know Minds*. Cambridge, MA: Harvard University Press.

Reddy, V. and Trevarthen, C. (2004). 'What We Learn about Babies from Engaging with their Emotions'. *Zero to Three* 24 (3) (January): 9–15.

Ridley, M. (2011). *Nature via Nurture*. London: Fourth Estate.

Rizzolatti, G. and Craighero, L. (2004). *Annual Review of Neuroscience* 7: 169–92.

Rizzolatti, G. and Fabbri-Destro, M. (2008). 'The Mirror Neuron System and Its Role in Social Cognition'. *Current Opinion in Neurobiology* 18 (2): 179–84.

Rizzolatti, G., Fogassi, L. and Gallese, V. (2001). 'Neurophysiological Mechanisms Underlying the Understanding and Imitation of Action'. *Nature Reviews Neuroscience* 2 (9): 661–70.

Rodier, P. M. and Arndt, T. L. (2005). 'The Brainstem in Autism'. In Bauman, M. L. and Kemper, T. L. (eds), *The Neurobiology of Autism*, 2nd edn. Baltimore, MD: Johns Hopkins University Press, 136–49.

Rogers, K. (2011). *The Brain and the Nervous System*. New York: Britannica Educational Publishing.

Rogers, P. J., Kainth, A. and Smit, H. J. (2001). 'A Drink of Water Can Improve or Impair Mental Performance Depending on Small Differences in Thirst'. *Appetite* 36: 57–8.

Rogoff, B. (2003). *The Cultural Nature of Human Development*. Oxford: Oxford University Press.

Rolls, E. T. (2004). 'The Functions of the Orbitofrontal Cortex'. *Brain and Cognition* 1: 11–29.

Rose, N. and Abi-Rached, M. (2013). *Neuro: The New Brain Science and the Management of the Mind*. Princeton, and Oxford: Princeton University Press.

Royal College of Midwives (2012). *Maternal Emotional Well-being and Infant Development: A Good Practice Guide for Midwives*. London: Royal College of Midwives Trust.

Rushton, S. (2011). 'Neuroscience, Early Childhood Education and Play: We are Doing It Right!' *Early Childhood Education Journal* 39: 89–94. DOI: 10.1007/s10643-011-0447-z (New York: Springer Science+Business Media, LLC).

Russo, S. J., Murrough, J. W., Han, M. H., Charney, D. S. and Nestler, E. J. (2012). 'Neurobiology of Resilience'. *Nature Neuroscience* 15: 1475–84.

Rutter, M. (1979). *Maternal Deprivation: New Findings, New Concepts, New Approaches*. London: Society for Research in Child Development Incorporated.

Sacher, J., Wilson, A. A., Houle, S., Rusjan, P., Hassan, S., Bloomfield, P. M., Stewart, D. E. and Meyer, J. H. (2010). 'Elevated Brain Monoamine Oxidase A Binding in the Early Postpartum Period'. *Archives of General Psychiatry* 67 (5): 468. DOI: 10.1001/archgenpsychiatry.2010.32.

Sander, L. W. (2008). *Living Systems, Evolving Consciousness and the Emerging Person: A Selection of Papers from the Life Work of Louis Sander*. Edited by Amadei G. and Bianchi I. New York and London: Analytic Press.

Sapolsky, R. M. (1996). 'Stress, Glucocorticoids, and Damage to the NS: The Current State of Confusion'. *Stress* 1: 1–19.

Satel, S. and Lilienfeld, S. O. (2013). *Brainwashed: The Seductive Appeal of Mindless Neuroscience*. New York: Basic Books.

Save the Children (2015). *Lighting Up Young Brains: How Parents, Carers and Nurseries Support Children's Brain Development in the First Five Years*. London: Save the Children.

Schaaf, R. C., Benevides, T., Blanche, E. I., Brett-Green, B. A., Burke, J. P., Cohn, E. S., Koomar, J., Lane, S. J., Miller, L. J., May-Benson, T. A., Parham, D., Reynolds, S. and Schoen, S. A. (2010). 'Parasympathetic Functions in Children with Sensory Processing Disorder'. *Frontiers in Integrative Neuroscience* 4: 4.

Schore, A. N. (2001). 'The Effects of Early Relational Trauma on Right Brain Development Affect Regulation and Infant Mental Health'. *Infant Mental Health Journal* 22 (1–2): 7–66, 201–69.

Schore, A. N. (2005). 'Attachment Affect Regulation and the Developing Right Brain: Linking Developmental Neuroscience to Paediatrics'. *Paediatrics in Review* 26: 204–11.

Schore, A. N. (2009). 'Attachment Trauma and the Developing Right Brain: Origins of Pathological Dissociation'. In Dell, P. F. and O'Neil, J. A. (eds), *Dissociation and the Dissociative Disorders DSM V and Beyond*. New York: Taylor and Francis Group.

Schore, J. R. and Schore, A. N. (2007). *Modern Attachment Theory: The Central Role of Affect Regulation in Development and Treatment*. Published online 8 September 2007. Berlin: Springer Science and Business Media, LLC.

Scientific Council on the Developing Child (SCDC) (2010). *Early Experiences Can Alter Gene Expression and Affect Long-Term Development*. Working paper no. 10, http://developingchild.harvard.edu/index.php/resources/reports_and_working_papers/working_papers/wp10 (accessed 5 June 2013).

Sethi, D., Bellis, M. A., Hughes, K., Gilbert, R., Mitis, F. and Galea, G. (2013). *European Report on Preventing Child Maltreatment*. Copenhagen: World Health Organization Regional Office for Europe.

Sherrington, C. S. (1955). *Man On His Nature*. Harmondsworth: Penguin Books Ltd. The Gifford Lectures, 1937–8. Cambridge: Cambridge University Press, 1940, 2nd rev. edn, 1951/53, Ch. 4, 'The Wisdom of the Body', 103–4.

Shonkoff, J. P. and Bales, S. N. (2011). 'Science Does Not Speak for Itself:

Translating Child Development Research for the Public and its Policymakers'. *Child Development* 82 (1): 17–32. DOI: 10.1111/j.1467-8624.2010.01538.x

Shonkoff, J. P. and Garner, A. S. (2012). 'Committee on Psychosocial Aspects of Child and Family Health; Committee on Early Childhood, Adoption, and Dependent Care; Section on Developmental and Behavioural Paediatrics. The Lifelong Effects of Early Childhood Adversity and Toxic Stress'. *Paediatrics* 129. DOI: e232–e246.

Shonkoff, J. P. and Phillips, D. (2000). *From Neurons to Neighbourhoods: The Science of Early Childhood*. Washington, DC: National Academic Press.

Shore, N. (1997). *Rethinking the Brain: New Insights into Early Development*. New York: Families and Work Institute.

Siegel, D. (2012). *The Developing Mind: How Relationships and the Brain Interact to Shape Who We Are*. New York: Guildford Press.

Sigman, M., Peña, M., Goldin, A. P. and Ribeiro, S. (2014). 'Neuroscience and Education: Prime Time to Build the Bridge'. *Nature Neuroscience* 17 (4): 497–502.

Siraj-Blatchford, I., Sylva, K., Muttock, S., Gilden, R. and Bell, D. (2002). *Researching Effective Pedagogy in the Early Years (REPEY), DfES Research Report 365*. London: DfES.

Spangler, G. Schieche, M., Ilg, U., Maier, U. and Ackermann C. (1994). 'Maternal Sensitivity as an Organizer for Biobehavioural Regulation in Infancy'. *Developmental Psychobiology* 27: 425–37.

Spelke, E. (1999). 'The Myth of the First Three Years: A New Understanding of Early Brain Development and Lifelong Learning'. *Nature* 401 (6754): 643–4.

Sperry, R. W. (1982). 'Some Effects of Disconnecting the Cerebral Hemispheres' (Nobel Lecture). *Science* 217: 1223–6.

Spitzer, M. (2012). 'Education and Neuroscience'. *Trends in Neuroscience and Education* 1 (1–2). DOI: 10.1016/j.tine.2012.09.002.

Sprenger, M. B. (2008). *The Developing Brain: Birth to Age Eight*. Thousand Oaks, CA: Corwin Press.

Sroufe, L. A. (1997). *Emotional Development: The Organization of Emotional Life in the Early Years*. Cambridge: Cambridge University Press.

Sroufe, L. A., Egland, B., Carlson, E. A. and Collins, W. A. (2005). *The Development of the Person: The Minnesota Study of Risk and Adaption from Birth to Adulthood*. New York: Guilford.

Stalnaker, T. A., Cooch, N. K. and Schoenbaum, G. (2015). 'What the Orbitofrontal Cortex Does Not Do'. *Nature Neuroscience*. 8: 620–27.

Stern, D. N. (2000). *The Interpersonal World of the Infant: A View from Psychoanalysis and Development Psychology*. 2nd edn. New York: Basic Books.

Stern, D. N. (2010). *Forms of Vitality: Exploring Dynamic Experience in Psychology, the Arts, Psychotherapy and Development*. Oxford: Oxford University Press.

Strathearn, L. (2006). 'Exploring the Neurobiology of Attachment'. In Mayes, L. C., Fonagy, P. and Target, M. (eds), *Developmental Science and Psychoanalysis: Integration and Innovation*. London: Karnac Press.

Stroganova, T. A., Orekhova, E. V. and Posikera, I. N. (1998). 'Externally and Internally Controlled Attention in Infants: An EEG study'. *International Journal of Psychophysiology*. 30: 339–51.

Sylva, K., Melhuish, E., Sammons, P., Siraj-Blatchford, I. and Taggart, B. (2004).

The Effective Provision of Pre-School Education (EPPE) Project: Final report – A Longitudinal Study Funded by the DfES 1997–2004. London: University of London, Institute of Education.

Taga, G. and Asakawa, K. (2007). 'Selectivity and Localization of Cortical Response to Auditory and Visual Stimulation in Awake Infants Aged 2 to 4 Months'. *NeuroImage* 36: 1246–52.

Tarullo, A. R. and Gunnar, R. (2006). 'Child Maltreatment and the Developing HPA Axis'. *Hormones and Behaviour* 50 (4): 632–9.

Teaching and Learning Research Programme (2007). *Neuroscience and Education: Issues and Opportunities. A Commentary by the Teaching and Learning Research Programme.* London: Institute of Education.

The Office of Communications (Ofcom) *Children and Parents: Media Use and Attitudes Report* (2015). London: Ofcom.

Thomas, M. S. C. and Knowland, V. C. P. (2009). *Sensitive Periods in Brain Development – Implications for Education Policy.* London: Birkbeck University.

Thompson, R. A. (2001). 'Sensitive Periods in Attachment?' In Bailey, D. B., Bruer, J. T., Symons, F. J. and Lichtman, J. W. (eds), *Critical Thinking about Critical Periods.* Baltimore, MD: Brooks Publishing Co., 83–106.

Thompson, R. A. (2006a). 'Conversation and Developing Understanding: Introduction to the Special Issue'. *Merrill-Palmer Quarterly* 52 (1): 1–16.

Thompson, R. A. and Goodvin, R. (2005). 'The Individual Child: Temperament Emotion, Self and Personality'. In Bornstein, M. H. and Lamb, M. E. (eds), *Developmental Science: An Advanced Textbook,* 5th edn. Mahwah, NJ: Lawrence Erlbaum Associates.

Thompson, R. A., Laible, D. J. and Ontai, L. L. (2003). 'Early Understanding of Emotion, Morality and the Self: Developing a Working Model', in Kail, R. V. (ed), *Advances in Child Development and Behaviour.* 31: 131–71.

Thouvenelle, S., Borunda, M. and McDowell, C. (1994). 'Replicating Inequities: Are we Doing it Again?' In Wright, J. and Shade, D. (eds), *Young Children: Active Learners in a Technological Age.* Washington, DC: National Association for the Education of Young Children.

Trevarthen C. (1978). 'Modes of Perceiving and Modes of Acting'. In Pick, H. L. Jr and Saltzman, E. (eds), *Psychological Modes of Perceiving and Processing Information.* Mahwah, NJ, and London: Lawrence Erlbaum Associates, 99–136.

Trevarthen C. (2000). 'Autism as a Neurodevelopmental Disorder Affecting Communication and Learning in Early Childhood: Prenatal Origins, Post-natal Course and Effective Educational Support'. *Prostaglandins Leucot: Essential Fatty Acids* 63: 41–6. DOI: 10.1054/plef.2000.0190.

Trevarthen, C. (2002). 'Learning in Companionship'. *Education in the North: The Journal of Scottish Education,* New Series, 10: 16–25. University of Aberdeen, Faculty of Education.

Trevarthen C. (2011). 'What is it Like to be a Person Who Knows Nothing? Defining the Active Intersubjective Mind of a New-born Human Being'. *Infant and Child Development* 20 (1): 119–35.

Trevarthen, C. (2011). 'What Young Children Give to their Learning, Making Education Work to Sustain a Community and its Culture'. *European Early Childhood Education Research Journal (Special Issue, 'Birth to Three',* Sylvie Rayna and Ferre Laevers, eds, 19 (2): 173–93. DOI: 10.1080/1350293X.2011.574405

Trevarthen, C. (2012). 'Finding a Place with Meaning in a Busy Human World: How Does the Story Begin, and Who Helps?' *European Early Childhood Education Research Journal* 20 (3): 303–12. doi:10.1080/135 0293X.2012.704757

Trevarthen, C. and Aitken, K. J. (2001). 'Infant Intersubjectivity: Research, Theory, and Clinical Applications'. *Journal of Child Psychology Psychiatry* 42 (1): 3–48.

Trevarthen, C., Aitken, K. J., Vandekerckhove, M., Delafield-Butt, J. and Nagy, E. (2006). 'Collaborative Regulations of Vitality in Early Childhood: Stress in Intimate Relationships and Postnatal Psychopathology'. In Cicchetti, D. and Cohen, D. J. (eds), *Developmental Psychopathology, Volume 2: Developmental Neuroscience*, 2nd edn. New York: Wiley, 65–126.

Trevarthen, C., Barr, I., Dunlop, A. W., Gjersoe, N., Marwick, H. and Stephen, C. (2003). *Review of Childcare and the Development of Children Aged 0–3: Research Evidence, and Implications for Out-of-Home Provision Supporting a Young Child's Needs for Care and Affection, Shared Meaning and a Social Place*. Edinburgh: Scottish Executive.

Trevarthen, C. and Bjørkvold, J.-R. (2016). 'Life for Learning: How a Young Child Seeks Joy with Companions in a Meaningful World'. Narvaez, D ., Braungart-Rieker, J., Miller, L., Gettler, L. and Hastings, P. (eds), *Contexts for Young Child Flourishing: Evolution, Family and Society*. New York: Oxford University Press, Ch. 2.

Trevarthen, C. and Delafield-Butt, J. (2013a). 'Autism as a Developmental Disorder in Intentional Movement and Affective Engagement'. *Frontiers in Integrative Neuroscience* 7: 49.

Trevarthen, C. and Delafield-Butt, J. (2013b). 'Biology of Shared Experience and Language Development: Regulations for the Inter-subjective Life of Narratives'. In Legerstee, M., Haley, D. W. and Bornstein, M. H. (eds), *The Infant Mind: Origins of the Social Brain*. New York: Guildford Press, 167–99.

Trevarthen, C. and Delafield-Butt, J. (2015). 'The Infant's Creative Vitality, in Projects of Self-discovery and Shared Meaning: How They Anticipate School, and Make it Fruitful'. In Robson, S. and Quinn, S. F. (eds), *The Routledge International Handbook of Young Children's Thinking and Understanding*. London: Routledge, 3–18.

Trevarthen, C., Gratier, M. and Osborne, N. (2014). 'The Human Nature of Culture and Education'. *Wiley Interdisciplinary Reviews: Cognitive Science* 5 (March/April): 173–92. *WIREs Cogn Sci* 2014. DOI: 10.1002/wcs.1276.

Trevarthen, C. and Reddy, V. (2007). 'Consciousness in Infants'. In Velman, M. and Schneider, S. (eds), *Companion to Consciousness*. Oxford: Blackwell, 41–57.

Tronick, E. Z. (1989). 'Emotions and Emotional Communication in Infants'. *American Psychologist* 44 (2): 112–19.

Twardosz, S. (2012). 'Effects of Experience on the Brain: The Role of Neuroscience in Early Development and Education'. *Early Education and Development* 23 (1): 96–119.

Twardosz, S. and Lukzker, J. (2010). 'Child Maltreatment and the Developing Brain: A Review of Neuroscience Perspectives', *Aggression and Violent Behaviour* 15 (1): 59–68.

United Nations Educational, Scientific and Cultural Organization (UNESCO)

(2012). *Technical Paper 9: A Place to Learn: Lessons from Research on Learning Environments*. Quebec: UNESCO Institute for Statistics.

United Nations Standing Committee on Nutrition (2010). *Scaling Up Nutrition: A Framework for Action*. New York: United Nations.

United Nations Educational, Scientific and Cultural Organization (UNESCO) (2007). *Education for All. Global monitoring report: Strong Foundations*. Paris: UNESCO.

Vygotsky, L. S. (1978). *Mind in Society: The Development of Higher Psychological Processes*. Edited by Cole M. , John-Steiner V., Scribner S. and Souberman E. Cambridge, MA: Harvard University Press.

Waddington, C. H. (1939). *An Introduction to Modern Genetics*. London: George Allen and Unwin Ltd.

Wave Trust Report http://www.wavetrust.org/our-work/publications/reports/1001-critical-days-importance-conception-age-two-period (accessed 1 September 2015).

Weaver, I. C., Cervoni, N., Champagne, F. A., D'Alessio, A. C., Sharma, S., Seckl, J. R., Dymov, S., Szyf, M. and Meaney, M. J. (2004). 'Epigenetic Programming by Maternal Behaviour'. *Nature Neuroscience* 7: 847–54 (Bethesda, MD: US National Library of Medicine).

Weisberg, D. S., Keil, F. C., Goodstein, J., Rawson, E., and Gray, J. R. (2007) 'The Seductive Allure of Neuroscience Explanations'. *Journal of Cognitive Neuroscience* 20: 470–7.

Werker, J. F. and Tees, R. C. (1984) 'Cross-language Speech Perception: Evidence for Perceptual Reorganization during the First Year of Life'. *Infant Behaviour and Development* 7: 49–63.

Werker, J. F. and Curtin, S. (2005). PRIMIR: 'A Developmental Framework of Infant Speech Processing'. *Language Learning and Development* 1 (2): 197–234.

West, M. and Prinz, J. (1987). 'Parental Alcoholism and Childhood Psychopathology'. *Psychological Bulletin* 102 (2): 204–18.

White, E. J., Hutka, S. A., Williams, L. J. and Moreno, S. (2013). 'Learning, Neural Plasticity and Sensitive Periods: Implications for Language Acquisition, Music Training and Transfer across the Lifespan'. *Frontiers in System Neuroscience*. Published online 20 November 2013. DOI: 10.3389/fnsys.2013.00090 (accessed 20 November 2014).

Wilkinson, R. and Pickett, K. (2010). *The Spirit Level*. London: Penguin Books.

Winnicott, D. W. (1988). *Babies and their Mothers*. London: Free Association Books.

Wolfe, P. (2007). *Mind, Memory and Learning: Translating Brain Research to Classroom Practices*. Napa Valley, CA: Association for Supervision and Curriculum Development (ASCD).

Wolfe, P. (2010). 'Brain Matters'. *Translating Brain Research into Classroom Practice*. Alexandria, VA: ASCD.

World Health Organization (WHO) (2014). *Investing in Children: The European Child Maltreatment Prevention Action Plan 2015–2020*. Denmark: World Health Organization.

Yelland, G. W., Pollard, J. and Mercuri, A. (1993). 'The Metalinguistic Benefits of Limited Contact with a Second Language'. *Applied Psycholinguistics* 14: 423–44.

Young, K. (1998). *Caught in the Net: How to Recognize the Signs of Internet Addiction – and a Winning Strategy for Recovery*. New York: John Wiley.

Yun, K. (2013). 'On the Same Wavelength: Face-to-Face Communication Increases Interpersonal Neural Synchronisation'. *Journal of Neuroscience* 33 (12): 5081–2.

Zeitlin, H. (1994). 'Children with Alcohol Misusing Parents'. *British Medical Bulletin* 50 (1): 139–51.

Zhu, H., Fan, Y., Guo, H., Dan Huang, D. and He, S. (2014). 'Reduced Interhemispheric Functional Connectivity of Children with Autism Spectrum Disorder: Evidence from Functional Near Infrared Spectroscopy Studies'. *Biomedical Optical Express* 5 (4): 1262–74.

Zimmerman, F. J., Christakis, D. A. and Meltzoff, A. N. (2007). 'Association between Media Viewing and Language Development in Children Under 2 Years'. *Journal of Paediatrics* 151: 354–68.

Websites

http://www.autism.org.uk/about.aspx (accessed 29 January 2016).

http://www.bbc.co.uk/news/health-244462922 (accessed 20 December 2014).

http://blogs.kent.ac.uk/parentingculturestudies/files/2011/09/Special-briefing-on-The-Myth.pdf (accessed 23 February 2015).

http://blogs.unicef.org/2014/09/20/neuroscience-is-redefining-early-childhood-development/ (accessed 20 November 2014).

http://www.cdc.gov/ncbddd/autism/data.html (accessed 29 January 2016).

http://www.drinkneuro.com/the-drinks/sleep (accessed 13th November 2014).

http://www.ehow.com/how_5011096_use-brain-gym-classroom.html (accessed 13 November 2014).

http://www.eif.org.uk/publication/social-and-emotional-learning-skills-for-life-and-work/ (accessed 9 February 2016).

http://www.escap.eu/research/peter-fonagy-and-the-undermining-of-old-ideas-on-personality-disorder (accessed 20 April 2015).

http://www.esrc.ac.uk/ESRCInfoCentre/about/CI/CP/research_publications/index27.aspx?ComponentId=6545&SourcePageId=6557 (accessed 17 November 2014).

http://faculty.washington.edu/chudler/neurok.html (accessed 15 November 2014).

http://www.theguardian.com/education/2014/apr/26/misused-neuroscience-defining-child-protection-policy (accessed 26 April 2014).

http://www.medicalsociologyonline.org/resources/MedSoc-2013-Conference-Posters/Oluduro-poster.pdf (accessed 18 December 2014).

http://www.news-medical.net/news/20160428/Scientists-build-semantic-atlas-to-show-how-human-brain-organizes-language.aspx (accessed 16 June 2016).

http://www.nidcd.nih.gov/health/voice/pages/specific-language-impairment.aspx (accessed 16 December 2014).

http://www.nxtbook.com/nxtbooks/zerotothree/201211/#/48/OnePage (accessed 20 December 2014).

http://www.parentingposttrauma.co.uk/blog/why-i-love-using-neuroscience-in-early-years-and-parenting-work (accessed 25 March 2015).

http://thephenomenalexperience.com/content/how-fast-is-your-brain (accessed 8
     November 2014).

https://www.psychologytoday.com/blog/i-got-mind-tell-you/201508/the-amygdala-
     is-not-the-brains-fear-center (accessed 15 March 2016).

http://www.sheffieldhealthyschools.co.uk/downloads/NurtureArticleForSENCO.pdf
     (accessed 10 April 2016).

http://stakeholders.ofcom.org.uk/binaries/research/media-literacy/children-
     parents-nov-15/childrens_parents_nov2015.pdf (accessed 19 December 2015).

http://www.talk4meaning.co.uk/ (accessed 16 June 2016).

http://www.teachthought.com/learning/research-based-when-a-lab-is-not-a-
     classroom/ (accessed 26 February 2015).

http://webspace.ship.edu/cgboer/limbicsystem.html (accessed 14 March 2016).

Index

The letter *f* after an entry indicates a page that includes a figure.